Scheduling Variable Capacity Resources for Sustainability

This book is the first to provide an overview of variable capacity scheduling and how it helps balance the growth of computing with its increasing environmental impact by creating more resilient and efficient systems that adapt to the dynamic nature of resource availability, optimizing utilization and minimizing disruptions.

The book covers all aspects of variable capacity scheduling from a technical and a societal perspective. It considers the new age of renewable power generation and how computing services can contribute to sustainability and grid decarbonization. Contributors investigate techniques that can be deployed for schedulers to cope with, or even benefit from, changes in the number of computing resources and the nature of the power sources. They survey emerging computing devices, such as edge servers, as alternatives to classical cloud computing platforms; they identify applications for monitoring energy and regulating power; and they investigate energy minimization and risk-aware scheduling in real-time systems, asking: 'How can a server with a variable processing speed schedule jobs with hard deadlines while minimizing its energy consumption?' The authors address the societal impact of computing, exploring how the social sciences play a critical role in solving the computing sustainability challenge and including a holistic analysis of current and future approaches, taking the whole life-cycle and rebound effects into account.

Presenting all the corresponding challenges and opportunities of variable capacity scheduling, this is an excellent introduction to the topic for students, researchers, computing scientists, or concerned citizens interested in sustainability.

Scheduling Variable Capacity Resources for Sustainability

Edited by

Anne Benoit, Andrew A. Chien, and Yves Robert

CRC Press
Taylor & Francis Group
Boca Raton London New York

CRC Press is an imprint of the
Taylor & Francis Group, an **informa** business

Designed cover image: *Concepts of Variable Capacity Scheduling*, created by Rajini Wijayawardana.

First edition published 2026
by CRC Press
2385 NW Executive Center Drive, Suite 320, Boca Raton FL 33431

and by CRC Press
4 Park Square, Milton Park, Abingdon, Oxon, OX14 4RN

CRC Press is an imprint of Taylor & Francis Group, LLC

Library of Congress Cataloging-in-Publication Data
Names: Benoit, Anne editor | Chien, Andrew Andai editor |
Robert, Yves editor
Title: Scheduling variable capacity resources for sustainability / edited
by Anne Benoit, Andrew A. Chien, and Yves Robert.
Description: First edition. | Boca Raton FL : CRC Press, 2026. | Includes
bibliographical references and index.
Identifiers: LCCN 2025042237 (print) | LCCN 2025042238 (ebook) | ISBN
9781032916293 paperback | ISBN 9781032916286 hardback | ISBN
9781003564270 ebook
Subjects: LCSH: Computer capacity--Planning | Computer scheduling | Data
centers--Energy conservation
Classification: LCC QA76.9.C63 S34 2026 (print) | LCC QA76.9.C63 (ebook)
LC record available at https://lccn.loc.gov/2025042237
LC ebook record available at https://lccn.loc.gov/2025042238

ISBN: 978-1-032-91628-6 (hbk)
ISBN: 978-1-032-91629-3 (pbk)
ISBN: 978-1-003-56427-0 (ebk)

DOI: 10.1201/9781003564270

Typeset in CMR12 font
by KnowledgeWorks Global Ltd.

This book has been prepared from camera-ready copy provided by the editors.

Contents

Editors

Anne Benoit is Full Professor in the Computer Science Laboratory LIP at ENS Lyon, France, and a Senior Member of Institut Universitaire de France. She is the Editor in Chief of Parco, and she has been Associate Editor in Chief of JPDC and Parco. She was the Chair of the IEEE CS TCPP (2020–2024), and she has chaired the program committee of several major conferences in her field, in particular SC, IPDPS, ICPP and HiPC. Her research interests include algorithm design and scheduling techniques for parallel and distributed platforms, with a focus on energy awareness and resilience.

Andrew A. Chien is the William Eckhardt Professor at the University of Chicago and a Senior Scientist at Argonne National Laboratory, and in prior roles has served at Intel Vice President of Research and Professor at UCSD and University of Illinois. Dr. Chien is a global research leader for sustainable computing, datacenters and power grids, parallel computing, computer architecture, clusters, resource scheduling, and cloud computing, and has received numerous awards for his research. Dr. Chien is an ACM, IEEE, and AAAS Fellow, and earned his PhD, MS, and BS from M.I.T.

Yves Robert is Full Professor in the Computer Science Laboratory LIP at ENS Lyon, France. He is the author of 7 books, 170 journal papers and 250 conference papers. He is the editor of 11 book proceedings and 13 journal special issues. He is the advisor of 38 PhD students. His main research interests are scheduling techniques and resilient algorithms for large-scale platforms. He is a Fellow of the IEEE. He has been awarded the 2014 IEEE TCSC Award for Excellence in Scalable Computing, the 2016 IEEE TCPP Outstanding Service Award, and the 2020 IEEE CS Charles Babbage Award.

Editors

Contributors

Jonatha Anselmi
University Grenoble Alpes, Inria, CNRS, Grenoble INP, LIG
Grenoble, France

Anne Benoit
LIP, ENS Lyon, CNRS, Inria, University Lyon1, and Institut Universitaire de France
Lyon, France

Anne Blavette
University of Rennes, ENS Rennes, CNRS, IETR lab
Rennes, France

Danilo Carastan-Santos
University Grenoble Alpes, CNRS, Inria, Grenoble INP, LIG
Grenoble, France

Christophe Cérin
University Sorbonne-Paris-Nord
Villetaneuse, France

Andrew A. Chien
University of Chicago and Argonne National Laboratory
Chicago, USA

Florina M. Ciorba
University of Basel
Basel, Switzerland

Bruno Gaujal
University Grenoble Alpes, Inria, CNRS, Grenoble INP, LIG
Grenoble, France

Alain Girault
University Grenoble Alpes, Inria, CNRS, Grenoble INP, LIG
Grenoble, France

Wedan Emmanuel Gnibga
University of Rennes, Inria, CNRS, IRISA
Rennes, France

Amina Guermouche
Bordeaux INP, University of Bordeaux, CNRS, Inria, LaBRI
Bordeaux, France

Thomas Hérault
Inria Center of Bordeaux University, LaBRI
Bordeaux, France

Piotr Luszczek
MIT Lincoln Lab, University of Tennessee
Knoxville, USA

Jean-Marc Nicod
Université Marie et Louis Pasteur, SUPMICROTECH, CNRS, institut FEMTO-ST
Besançon, France

Anne-Cécile Orgerie
University of Rennes, Inria, CNRS, IRISA
Rennes, France

Stéphan Plassart
University Savoie Mont-Blanc
Annecy, France

Yves Robert
LIP, ENS Lyon, CNRS, Inria, University Lyon1
Lyon, France

Pierre Segonne
Electricity Maps

Denis Trystram
University Grenoble Alpes, CNRS, Inria, Grenoble INP, LIG
Grenoble, France

Chapter 1

The Emerging Challenge of Variable Capacity Scheduling

Anne Benoit, Andrew A. Chien, and Yves Robert

This book focuses on managing and allocating resources in the new age of renewable power generation. High-performance computing platforms, datacenters and cloud computing environments all experience fluctuating capacity over time due to external factors like weather patterns or grid conditions. From a computing perspective, the goal is to optimize resource utilization and minimize disruptions by dynamically adjusting job scheduling and resource allocation strategies in response to these capacity variations. From a societal perspective, the opportunity is to enhance sustainability efforts and to minimize the environmental impact of computing.

This introductory chapter details the motivation for variable capacity scheduling, provides a formal definition of the problem and discusses new challenges and opportunities, together with societal impact. All these topics are further discussed in the following chapters, whose organization we describe in the last section.

1.1 Prologue

Humankind faces unprecedented environmental challenges. Since the Industrial Revolution on the European continent in the mid-19th century, human activities have led to a major ecological crisis (global warming, species extinction, depletion of natural resources such as water and metals) [9].

We have hence gathered the research community focused on compute scheduling in the new age of renewable power generation – where variation in weather and solar radiation drives renewable generation, and thereby the carbon intensity of power, and efforts to reduce environmental damage then

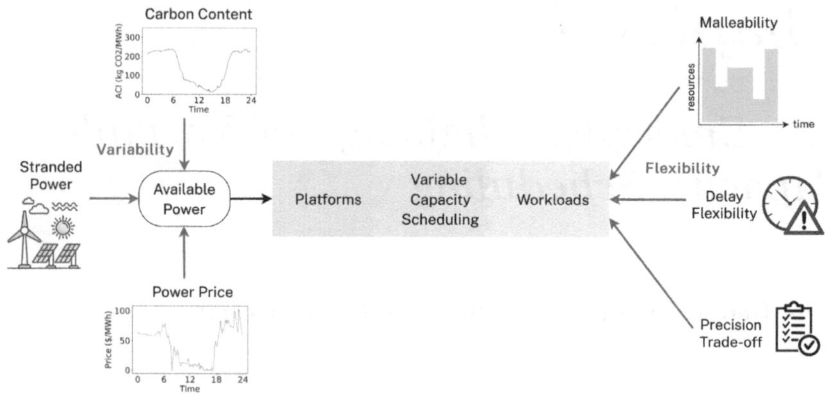

Figure 1.1: The Variable Capacity Scheduling challenge includes dimension of platforms (power sources and carbon emissions) and workflows (flexibility, malleability, precision).

create variation in compute capacity (opportunity). In this age, effective utilization of that opportunity is the key to sustainable computing. Discussions included important findings on the attributes of future platforms, characterizing the power grid variation, reliability challenges, and two-way relationships between datacenters and grids that are emerging. Workloads are changing as well, with growing understanding of how computing can be malleable, delay flexible, or even acceptably approximate. Discussions also included how scheduling can meet these challenges, managing between flexible workloads and dynamically varying platforms – performing to new metrics that reflect both performance and non-performance attributes (e.g., carbon emissions). The concerns that underlie a shift to variable capacity derive from a broader societal concern about the sustainability of computing, so we also address critical challenges in awareness, responsibilities, and action. These multi-faceted concerns are depicted in Figure 1.1.

1.2 Why Variable Capacity Datacenters Are Important for Sustainability

Over the past 25 years, the world has achieved a significant transformation of its electric power generation. From nearly 20% in 2000 (mostly hydroelectric), today's global electricity generation (2024) is nearly 30% renewable, with most of the increase coming from new wind and solar power generation. In the next 25 years, this fraction is expected to increase to over 75% renewables (by

2050). But progress is extremely varied, with some localities already reaching over 75% today (e.g., Norway, Canada, Brazil), others approaching 50% (e.g., California, Germany), slow starters (e.g., Japan, USA, India, China) and others nearly at the starting line (e.g., South Korea, Russia, Pakistan, Iran).

However, there is a major mismatch between new generation added to the grid – mostly variable renewables such as wind and solar – and the needs of datacenters for continuous reliable power. This "balancing" of generation and load was traditionally achieved by power grid dispatch of fossil-fuel generation. The ability to dictate generation level is a key element of this model, but it has been disrupted. Volatile renewables generate whenever the sun or wind is available and cannot be commanded to increase their generation. To solve this new problem, there are two major approaches: 1) energy storage, and 2) adaptive load shaping. While storage costs continue to decrease, the low price of energy in many markets means that balancing purely with energy storage (daily, weekly, monthly, and annually) remains prohibitively expensive. The opportunity for shaping load is a new grid need, and hence an opportunity. These are variable and adaptive loads [2, 4, 5], as they will not only ease balancing difficulty, but also increase renewable absorption.

1.3 Static to Dynamic Datacenter Management

Traditional datacenter design has targeted fixed capacity operation – which could be ensured 24/7. With a target information technology (IT) equipment power, design for power distribution and cooling was optimized for cost and energy efficiency with the most popular metrics being Total-Cost-of-Ownership (TCO) and power-use efficiency (PUE) [1, 3].

This paradigm produced significant advances but was incomplete in several dimensions, causing losses in *real efficiency* not captured by the metrics. The traditional paradigm failed to capture variation in compute load, external weather, and seasonal variation. And of course, it failed to capture power grid variations, which are a focus of this book. These issues have been recognized by the research community [3, 6] and can easily produce unexploited capacity of 20-50%. For example, daily ambient temperature variations (day-night, 15 degrees Celsius in many locations) can affect cooling power by 100% and datacenter capacity by up to 100%. Likewise, seasonal variations (winter-summer, 30 degrees Celsius average in many locations) can produce similar changes in cooling capacity. Daily and seasonal variation in grid generation (e.g., hydro, wind, solar) and power load can cause large variations in power price (200-400%) and carbon emissions (0-100%).

Furthermore, many datacenters experience diurnal variation in load reducing their average utilization to 40-60%, causing them to fall far short of their idealistic TCO and PUE efficiency analyses. These shortfalls can range from 20% to 50%.

To recognize these omissions, advanced research and datacenter design employ dynamic metrics for capacity and have begun to optimize for operation across that entire range and even for local statistics [3, 6, 12]. This creates the foundation for scheduling of variable capacity resources.

Benefits of Dynamic Datacenters for the Power Grids. Beyond increasing datacenter capacity and enabling lower costs, dynamic datacenters have benefits for the power grid. By increasing power demand when there is excess renewable energy, datacenters can help the grids absorb more carbon-free energy. This helps the power grid decarbonize and further helps solve the balancing problem between load and generation. By decreasing power demand when there is shortfall of renewable energy, datacenters reduce the need for grid energy storage and transmission, reducing grid costs, and thereby power costs. If instead the shortfall would be covered by running fossil-fuel generation, then timely datacenter power reduction would reduce grid carbon emissions. To support all of these scenarios, datacenters must become dynamic, responding to their external – weather, grid, and more – environments to adapt their operating capacity. This produces a "scheduling for variable capacity resources" problem.

1.4 Variable Capacity Scheduling Problem

1.4.1 Problem statement

With growing global concern about climate change [14, 15], the end of Dennard scaling, and continued exponential growth of computing use [10, 13], there is growing interest in how to reduce the negative environmental impacts of computing, that is, improve its sustainability [8, 11, 16, 17]. A popular approach that exploits the variation of renewable generation (wind, solar) is the idea of temporal and spatial load shifting to match computing power consumption with the availability of low-carbon power [7]. Viewed at a single computing site, this appears as a datacenter or high-performance computing (HPC) platform whose capacity evolves with time, depending upon the cost and the environmental policy (e.g., reduce power supply when the power source is coal instead of wind or solar), termed the *variable capacity scheduling* problem [19].

We formally state the *variable capacity scheduling* problem as follows. In a datacenter or cluster, let M denote the number of total machines, where each machine m has $r(m)$ resources. We want to schedule a set of jobs J on the M machines. Each job $j \in J$ has submission time $s(j)$, resource requirement $r(j)$, and execution time $t(j)$. The scheduler needs to decide $j_{m,t}$, which is the decision variable of running job j on machine m at time t. In traditional systems, such placements are subject to each machine resource constraint:

$$\forall t, \forall m \in M, \quad \textstyle\sum_{j \in J} j_{m,t} r(j) \leq r(m),$$

where the left-hand side calculates the number of active resources that are processing jobs on each machine. However, in the new scheduling problem with resource capacity variations, the overall available resource capacity is a function of time t. It is denoted as $R(t)$, where $R(t) \leq M$. Hence, all job placements are now subject to a time-varying resource capacity constraint at each time-slot t:

$$\begin{array}{l} \forall t, \forall m \in M, \quad \sum_{j \in J} j_{m,t} r(j) \leq r(m) \\ \text{subject to} \\ \forall t, \forall m \in M, \quad u_{m,t} = 1 \quad \Leftrightarrow \quad \exists j \in J \quad s.t. \quad j_{m,t} = 1, \\ \forall t, \quad \sum_{m \in M} u_{m,t} r(m) \leq R(t). \end{array}$$

These constraints ensure that the total number of machines that have active running jobs do not exceed current resource capacity $R(t)$, where $u_{m,t}$ indicates whether a machine is active or not. In this formulation, the resource capacity $R(t)$ may either correspond to a number of available machines at time t, or to a budget fo green power at time t, and so on. It is easy to account for an extra cost paid when exceeding resource capacity instead of enforcing a strict limit. Chapter 3 provides a detailed description of scheduling techniques that implement temporal and spatial load shifting to match resource variability on the fly.

It is useful to define some common terminology to describe the *variation* of resource capacity [18, 19]. Specifically, we define

- **Dynamic Range:** the difference between the maximum and minimum capacity of the computing resources, corresponding to $\max_t\{R(t)\} - \min_t\{R(t)\}$.

- **Change Frequency:** the frequency of capacity changes, such as daily, hourly, every 15 minutes, every minute, etc. $R(t)$ remains piece-wise constant between two capacity changes.

- **Variability Structure:** the statistical type of variation such as random walk, uniform random, etc. which defines how much the capacity changes within the dynamic range, at each change.

- **Predictability** (or foresight [19]): the accuracy with which the capacity can be predicted, and how far in advance.

These four key characteristics (dynamic range, change frequency, variability structure, and predictability) frame the variable capacity scheduling problem as distinct from classical scheduling challenges.

1.4.2 New challenges and opportunities

Platforms. The greatest challenge at the platform level is to describe resource capacity as a function of time, and to assess how accurately it can be forecast. At the grid level, a key question is, how can capacity planning be determined and coordinated across different autonomous systems (with different incentives and objectives) for better overall outcomes? Opportunities to increase the use of green power sources and decrease carbon emissions are reinforced by an effective management of the constraints induced by data and computation movement inside each platform and across them.

Workloads. An important issue is to extend workload specifications to incorporate flexibility. In addition to promoting moldable and malleable jobs over rigid jobs, it is crucial to consider different categories of workload urgency and timing, such as mandatory and instant, mandatory and non-instant, and non-mandatory and (non-)instant. These categories provide great opportunities for the batch scheduler to shift load in time and space, but a related question is to give incentives for users to accept some delays and a (hopefully small) performance degradation.

Scheduling Models and Metrics. As for models, both workloads and platforms encompass high degrees of uncertainty. A workload with uncertain requirements may lead to overestimating or underestimating resource requirements. Platforms with high capacity variations may lead to interrupting or reshaping applications. Uncertainty must be accounted for globally by scheduling and resource management algorithms. As for metrics, a single objective optimization function is no longer enough; we should consider multiple objectives at a time, e.g., performance capping and energy minimization. Furthermore, how do we expose and exchange information between jobs and schedulers to enable cooperation? What is the interplay between the user-facing incentives and the job scheduling objectives (which include delays, carbon-saving, migration, etc.)? Do we need pricing models and negotiations to model and study this interplay? Finally, as for scheduling algorithms: what are the scheduling techniques that allow us to move from single platforms to distributed platforms, to allow us to exploit workload flexibility and variable resource capacity?

1.5 Societal Impact

We are not living in a world with infinite resources. Our community of computer scientists are aware of the urgency of the climate warming situation, and we all (especially computer scientists) must face the reality of finite global computing capacity. This means we limit the growth of IT, particularly difficult in the face of artificial intelligence's explosive growth into all sectors of society. The critical limits include not only power, but also the availability of materials necessary to build digital components for computing.

Hence, an important step is to identify and classify existing mechanisms that could help users to accept the constraints linked to environmental rules; clearly, preserving fairness among all users is the only way common people will accept the constraints. Such dramatic change from our current frame will take a long time to achieve, although there is an urgent need to move on. The effort should be collective, and we will not be able to succeed without the help of colleagues coming from other disciplines, including sociologists, economists, and philosophers.

1.6 Organization of Remainder of Book

This book is organized into four parts. The first part discusses the variable capacity scheduling problem in full details. Chapter 2 focuses on datacenters, describing how new power distribution strategies create opportunities for dynamic flexibility in capacity, the key element required for grid integration, and the driving motivation for variable capacity scheduling. The chapter provides a full discussion of the important role that datacenters can play in both increasing grid power capacity and accelerating grid decarbonization. This chapter surveys the impact of resource variability onto scheduling algorithms and metrics. Chapter 3 investigates several scheduling techniques that can be deployed for batch schedulers to cope with, or even benefit from, changes in the number of computing resources and the nature of the power sources. The chapter continues with a broad overview of the challenges that face scheduling techniques to account for variability, from new workload types up to combined optimization metrics.

The second part of the book deals with infrastructures and applications. Chapter 4 surveys emerging computing devices such as edge servers and shows that these distributed low-cost resources can now run powerful applications efficiently, thereby becoming an alternative to classical cloud computing platforms. However, their success has a high environmental price and the chapter focuses on building efficient and sustainable systems, surveying a wide spectrum of strategies to optimize infrastructure operation. The chapter concludes with a discussion on cooling systems and their interplay with energy management techniques for datacenters. Chapter 5 focuses on sustainable computing and presents a comprehensive analysis of how computing services can contribute to grid decarbonization. The chapter critically evaluates carbon-aware computing and proposes a novel four-step blueprint for sustainable computing that prioritizes actions based on their verified impact on emissions reduction.

The third part of the book considers power optimization for job execution. Chapter 6 identifies key application characteristics and uses them to monitor energy, regulate power and improve scheduling. Techniques like DVFS, core frequency scaling and uncore frequency scaling are explained in detail. Then, a detailed experimental setup is presented to demonstrate the impact of these techniques. The chapter continues with a study of power capping and its impact on consumption and performance. Chapter 7 investigates energy minimization in real-time systems: How can a server with a variable processing speed schedule jobs with hard deadlines while minimizing its energy consumption? The chapter uses an approach with continuous processor speeds and considers several problems of increasing difficulty, first in a deterministic setting, and then in a stochastic setting. The chapter provides background, scheduling algorithms and complexity results (with proofs only for new results), and the results range from worst-case guarantees to statistical learning.

The fourth part of the book addresses the societal impact of computing. Chapter 8 begins with a discussion of the environmental impacts of HPC platforms and datacenters, which significantly contribute to carbon emissions. It next considers how the goals of HPC (performance alone) can be enhanced with "green" metrics, the consequence of which can be to reshape applications and their use. Finally, it extends considerations of the direct impact of the HPC to the secondary and tertiary consequences in terms of environmental impact – of the computations and even the scientific as well as societal follow-on impact of the computing results. This is framed in terms of a consequential life-cycle analysis, which can be used to holistically assess and perhaps limit future computing.

1.7 Acknowledgements

This volume came out of a workshop held on March 29-31, 2023 in Paris, France at the University of Chicago Paris Center.

The talks and position papers from attendees are available at `https://people.cs.uchicago.edu/~aachien/workshops/varsched23/`.

We thank all workshop participants, with a special mention to Florina Ciorba and Denis Trystram, who contributed to the workshop report, and to our students Svetlana Kulagina, Lucas Perotin, and Rajini Wijayawardana, who acted as recorders during all sessions and group discussions.

Finally, we thank the International Institute of Research in Paris (IIRP) of the University of Chicago for their financial support, for hosting the workshop, and for their help with its organization. We thank the FACCTS and CNRS-UChicago programs for enabling a fruitful collaboration between the University of Chicago and Laboratoire LIP at École Normale Supérieure de Lyon.

1.8 References

[1] L. A. Barroso, U. Holzle, P. Ranganathan, and M. Martonosi. *The Datacenter As a Computer: Designing Warehouse-scale Machines.* Morgan & Claypool Publishers, 3rd edition, 2018.

[2] CAISO. CAISO Market Data. `https://www.caiso.com/todays-outlook/`.

[3] A. A. Chien, C. Zhang, and L. Lin. Beyond PUE: Flexible datacenters empowering the cloud to decarbonize. *USENIX Hot Carbon*, 2022.

[4] B. F. Gerke, G. Gallo, S. J. Smith, J. Liu, S. V. Raghavan, P. Schwartz, M. A. Piette, R. Yin, and S. Stensson. The California Demand Response Potential Study, Phase 3: Final Report on the Shift Resource through 2030, 07/2020 2020.

[5] B. F. Gerke, S. J. Smith, S. Murthy, S. Baik, S. Agarwal, P. Alstone, A. Khandekar, C. Zhang, R. E. Brown, J. Liu, and M. A. Piette. The California Demand Response Potential Study, Phase 4: Report on Shed and Shift Resources Through 2050, 05/2024 2024.

[6] W. E. Gnibga, A. A. Chien, A. Blavette, and A. C. Orgerie. FlexCoolDC: Datacenter Cooling Flexibility for Harmonizing Water, Energy, Carbon, and Cost Trade-offs. In *Proc. of the 15th ACM Int. Conf. on Future and Sustainable Energy Systems*, e-Energy '24, page 108–122, 2024.

[7] Green ICT. Follow the sun, wind, moon, 2009. `https://www.vertatique.com/cloud-computing-starting-follow-sunwindmoon`.

[8] U. Gupta, M. Elgamal, G. Hills, G.-Y. Wei, H.-H. S. Lee, D. Brooks, and C.-J. Wu. ACT: Designing Sustainable Computer Systems with an Architectural Carbon Modeling Tool. In *49th Int. Symp. Computer Architecture (ISCA)*, 2022.

[9] Intergovernmental Panel on Climate Change (IPCC). *Global Warming of 1.5°C*, page 1–24. Cambridge University Press, 2022.

[10] N. Jones. How to stop data centres from gobbling up the world's electricity. *Nature*, September 2018.

[11] A. Radovanovic. Our data centers now work harder when the sun shines and wind blows, Apr 2020. `https://blog.google/inside-google/infrastructure/data-centers-work-harder-sun-shines-wind-blows`.

[12] Sidewalk Infrastructure Partners (SIP). Powering the data future sustainably: Introducing Verrus, March 2024. `https://sidewalkinfra.com/news/powering-the-data-future-sustainably-introducing-verrus`.

[13] Tal Rosenberg. The computer that will change everything. *Chicago Magazine*, January 2023. https://www.chicagomag.com/chicago-magazine/february-2023/the-computer-that-will-change-everything.

[14] United Nations Framework Convention on Climate Change. Kyoto protocol, 1997. http://unfccc.int/kyoto_protocol/items/2830.php.

[15] United Nations Framework Convention on Climate Change. Paris climate change conference, 2015. http://unfccc.int/meetings/paris_nov_2015/meeting/8926.php.

[16] United States National Science Foundation. Design for environmental sustainability in computing (desc), December 2022. https://beta.nsf.gov/funding/opportunities/design-environmental-sustainability-computing-desc.

[17] F. Yang and A. A. Chien. Large-scale and extreme-scale computing with stranded green power: Opportunities and costs. *IEEE Transactions on Parallel and Distributed Systems*, 29(5), December 2017.

[18] C. Zhang. *Eliminating the Capacity Variation Penalty for Cloud Resource Management.* PhD thesis, Department of Computer Science, University of Chicago, January 2023. https://newtraell.cs.uchicago.edu/research/publications/techreports/TR-2023-02.

[19] C. Zhang and A. A. Chien. Scheduling challenges for variable capacity resources. In *Job Scheduling Strategies for Parallel Processing: 24th International Workshop, JSSPP 2021, Virtual Event, May 21, 2021, Revised Selected Papers 24*, pages 190–209. Springer, 2021.

Part I

Variable Capacity Resource Problem

Chapter 2

From High-performance Computing Datacenters to Grid-integrated AI Datacenters

Andrew A. Chien

To provide understanding of the opportunities and challenges of datacenters, we describe the basic elements of a datacenter – including power distribution, cooling, and of course computing. We elucidate how these systems create opportunities for dynamic flexibility in capacity, the key element required for grid integration, and the driving motivation for variable capacity scheduling. We discuss grid dynamics and how growing datacenter demand challenges the decarbonization of the grid with variable renewable generation. This leads to discussion of the important role that datacenters can play in both increasing grid power capacity and accelerating grid decarbonization, and what datacenters (and computation workloads) need to do to help society go forward into a more sustainable future, combined with the growth of artificial intelligence. Finally, we discuss some implications for scheduling problems, and some interesting research directions.

2.1 Datacenter Basics

2.1.1 Elements of a datacenter

A modern datacenter has three key elements: (1) computing (IT) equipment, (2) power distribution, and (3) cooling (heat removal). These three parts are illustrated in Figure 2.1. While computer scientists often focus on the computing equipment, that limited view is not sufficient to characterize the datacenter capacity, as power/cooling limits determine the operating point – and therefore the capacity – of the compute equipment. A simple, but more complete

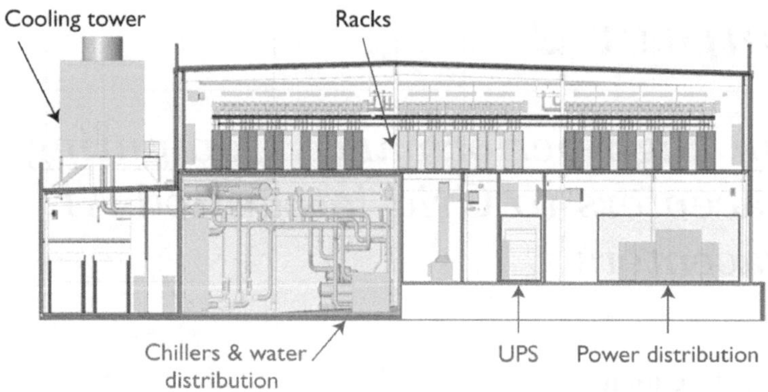

Figure 2.1: Datacenter elements: Computing equipment, power distribution, and cooling. UPS stands for the Uninterruptible Power Supply.

model of datacenter capacity would be:

$$DC_C = min(IT_C, Power_C, Cooling_C),$$

where each term is the respective capacity in megawatts. Hence, IT_C is the maximum power the computing can sustain – unconstrained by power distribution and cooling, $Power_C$ is the maximum power the power distribution system can sustain, and $Cooling_C$ is the maximum cooling capacity. This overall datacenter capacity constraint arises because the power must be supplied to enable the computing, and any power supplied must be removed as heat (cooling). Of course, the cooling capacity, $Cooling_C$ is really a function of time, so a more complete version of the equation is:

$$DC_C(t) = min(IT_C, Power_C, Cooling_C(t)).$$

In the basic static operation mode, the variation over time is often ignored (the minimum values are used), so for a given IT power, the power distribution and cooling are optimized for energy efficiency (e.g., minimum PUE). An additional consideration is sufficient redundancy for reliability. This produces infrastructure with excess capacity for power distribution and cooling. For example, for reliability, a redundancy of at least $N + 1$ is required in power distribution [57]. Furthermore, internal to the cooling system, flow rates and water temperatures chosen for energy leave ample opportunity for increased cooling capacity.

Opportunities for Dynamic Capacity. Datacenters are typically built in a fashion that creates multiple dimensions of opportunity for dynamic capacity. This includes all three dimensions: power distribution, computing, and cooling elements. We outline this below:

- Power distribution is built with redundancy for reliability of $2N$ (full redundancy), or at least $N+1$ (a spare). In this case, N determines the

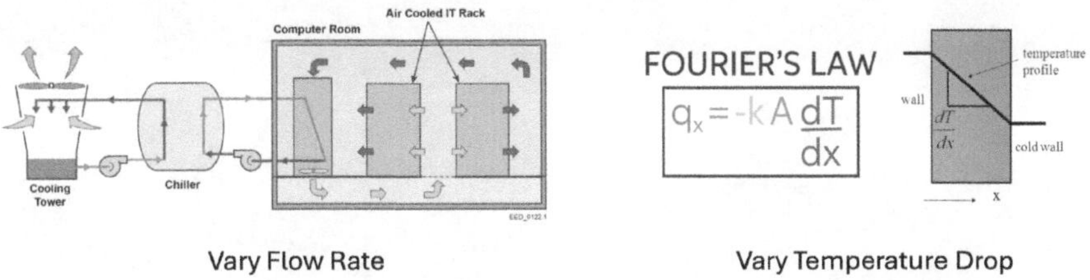

Figure 2.2: Increasing the flow rate or the temperature difference can increase the capacity of the cooling system.

power capacity, $Power_C$, so a $2N$ design can have 100% headroom for power, and less redundant $N + 1$ design might have 25-50% headroom. There might be a modest cost to exploit this headroom in increased power distribution interconnects.

- Computing – Many chips have turbo clock rates and multiple frequencies of operation. These turbo clock rates can be 1.5-3x or even more. Memory and storage chips have lesser speed flexibility, so system performance/capacity increases are likely to be smaller. The ability of modern chips to operate at these higher speeds is generally limited primarily by the junction temperatures of transistors on the chip. That in turn depends primarily only on the available cooling – infrastructure in heat sinks, piping – and current dynamic capacity.
- Cooling – Modern datacenters are water-cooled typically with water piped to the chips (CPUs, GPUs) and air flow for the residual. Cooling capacity can be increased with flow rate or lower temperature (see Figure 2.2). With estimates of 25% headroom in each of these dimensions, a 50% overall cooling headroom based on internal management is possible. And all datacenters have greater cooling capacity when the external temperature is lower; this headroom could exceed 100% (see Figure 2.3).

For example, dropping the water temperature by 5 degrees Celsius, or increasing the flow rate by 25% might increase the cooling capacity by 50%, enabling the compute hardware to be run faster [14, 23]. Of course, this comes with some penalty in the energy cost of the cooling [23] (see Figure 2.2). This perspective only covers the inside of the datacenter.

Moving outside the datacenter, the external environment has a major impact on the ability to cool, $Cooling_C$. Obviously, the external environment's temperature, where the heat is vented by the cooling system, has a big impact on how much heat can be removed (see Fourier's law). But other dimensions such as humidity, whether sunlight is falling on the datacenter, etc. can affect the total heat balance and thus can affect datacenter capacity (see Figure 2.3). Dimensions of sunlight, temperature, and humidity vary with weather and season dynamically. And sunlight in particular, away from the equator, has

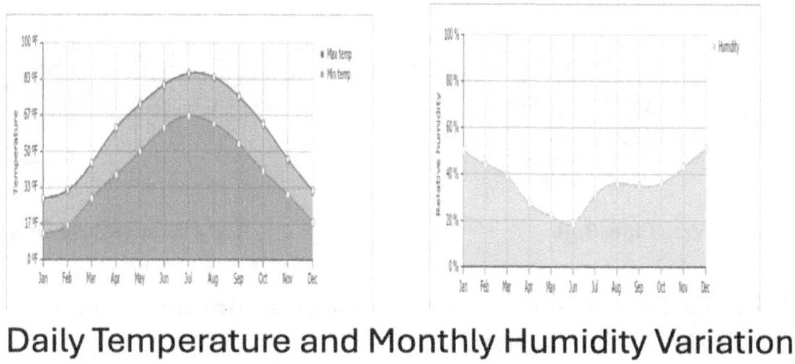

Figure 2.3: External temperature and humidity can affect the datacenter capacity by increasing the temperature drop, or the capacity of evaporative coolers.

a major temporal structure change across the seasons (e.g., shorter days in winter).

Overall, the internal adjustments can enable up to 30% increase in capacity, and with external environmentals, and even larger degree, as much as 50%. Of course, flexing down below 100% is easily achieved in nearly all datacenters.

2.1.2 AI datacenters: 10x power density and load

AI datacenters, a loose term for datacenters that support predominantly AI training and inference workload, are more demanding than conventional cloud datacenters in two key dimensions. At a technical level, their power density (kW/sqft) and total power (MW) are both typically 10-fold or higher due to large quantities of GPUs. The driver for this is that at a microscopic level, generative AI inference required for a service such as ChatGPT is 10-100x energy per request compared to conventional internet services, and at a macroscopic level, the use of generative AI is proliferating rapidly [43]. At an economic level, they are correspondingly more expensive. Thus, these large datacenters are driving grid power load growth, adding hundreds of megawatts or even gigawatts per datacenter complex, and must become adaptive loads [5, 6, 26, 43, 44].

2.1.3 Examples of modern datacenters

Today's datacenters already represent a large fraction of total grid load in some utilities such as Dominion Energy (24%, 2023) and EirGrid (15%, 2023) [30], and they are growing by as much as 25% per year. To calibrate for typical load and densities in today's datacenters, we consider two high-performance computing (HPC) datacenters, and two recent leading edge AI datacenters (see Figure 2.4).

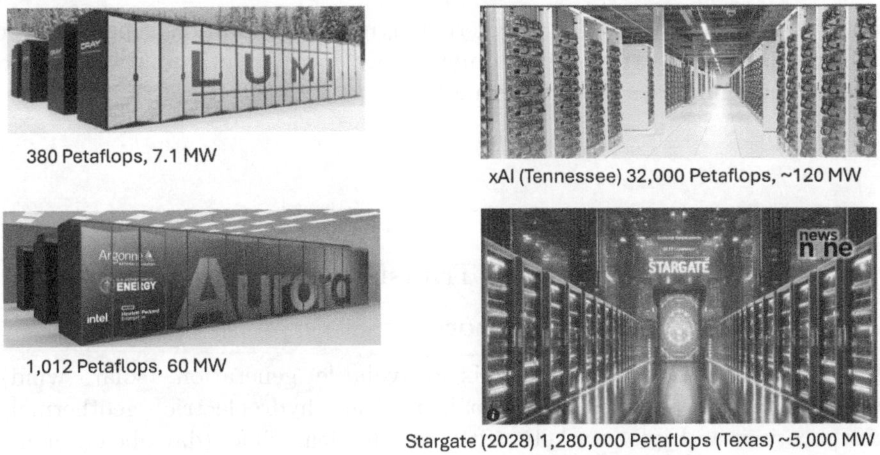

Figure 2.4: A range of modern datacenters: Capacity and potential dynamic adaptation range.

The world's third largest HPC system is Argonne's Aurora system [2]. The Aurora system consumes 54 MW of power, has 10,624 nodes, and with two CPUs per node, 21,248 CPUs, and with 6 GPUs per node, 63,744 GPUs total. With this hardware and a Cray Slingshot interconnect, the Aurora system achieves 2 exaflops [48]. The datacenter housing Aurora is approximately 100 MW capacity with 18,000 sqft of floor space, of which Aurora occupies 10,000 sqft. A 30% dynamic adaptation for Aurora's datacenter would be a range of 60 MW.

Recently, the Europeans created a new HPC datacenter designed for sustainability. This center is called the Large-Unified Modern Infrastructure (LUMI), and it is designed to support HPC and AI "factories". LUMI's first supercomputer is the world's eighth fastest, with 2,978 GPU nodes, each node with one 64-core AMD Trento CPU and four AMD MI250X GPUs [32]. This gives a total of 11,912 AMD GPUs. The LUMI system consumes 7.1 MW, and the facility is powered wholly by renewable energy and the waste heat is recycled. LUMI's first system occupies 400m^2 or approximately 4,000 sqft of floor space. The datacenter is planned to be much larger, perhaps 30 MW. A 30% dynamic adaptation for the LUMI datacenter would be in the range of 18 MW.

A recently constructed AI datacenter belonging to Elon Musk's xAI startup recently (November 2024) came online in Tennessee [54]. This system (Gigafactory of Compute) has 100,000 latest model GPUs, and a datacenter power specification of 150 MW. A 30% dynamic range for the xAI datacenter at its current size would be in the range of 90 MW.

And the largest, most ambitious datacenter to date is Stargate, announced on January 23, 2025 as a 5 GW facility. Few other details are available, but such a system might have a staggering dynamic adaptation range of over a gigawatt. That is truly extraordinary, both for computing as well as the benefits and challenges it would bring to the power grid.

2.2 Today's Power Grid: Transformation in Progress

Markets and volatile generators

The modern power grid is a mix of volatile generation (solar, wind) and traditional resources (gas, coal, nuclear, hydroelectric, geothermal) incorporating multiple markets with varying lead-time (day-ahead, hour-ahead, 15-minute, and real-time) and global economic dispatch to minimize grid power cost [16, 34]. Cited by the US National Academy of Engineering as a greatest engineering wonder of the 20th century [35], the power grid orchestrates generation resources (reliable and unreliable), and transmission to meet demand (load). The orchestrating (dispatch) uses markets to provide fair access, under the constraint of achieving reliable power. The economic dispatch is designed to also provide generally low-cost power. In Figure 2.5, we illustrate variation in renewable generation and load over months, weeks, and hours. Renewable generation in CAISO fluctuates more than 10-fold in each 24-hour period with significant differences in weekday and weekend patterns. Each grid is different, with dynamics

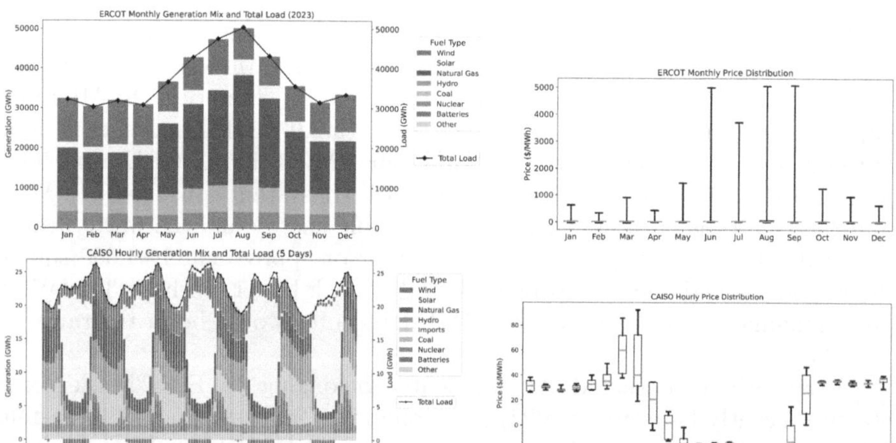

Figure 2.5: ERCOT generation mix and price variation for 12 months (2023), CAISO 5-day generation mix and price variation (April 2024). Grid behavior varies widely over monthly, weekly, daily, and hourly time scales.

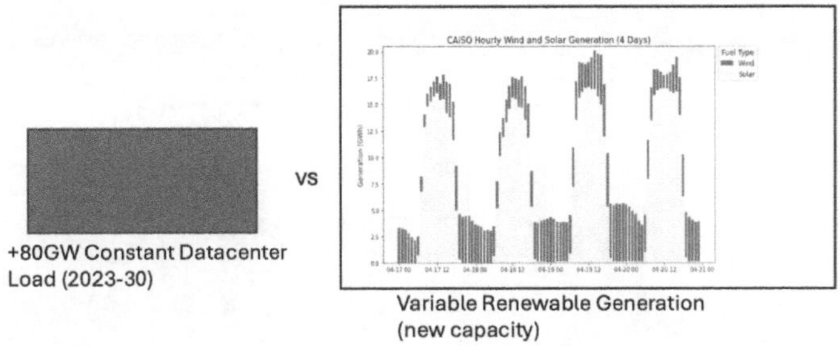

Figure 2.6: To balance the growth in Cloud/AI datacenter demand at gigawatt scale with Variable Renewable Generation is challenging.

produced by its specific mix of generation and load. For example, ERCOT's mix includes significant wind and thus exhibits a range of longer temporal structures. Both CAISO and ERCOT exhibit large seasonal variation. Power prices are even more dynamic, with variation from +\$40/MWh to -\$40/MWh in CAISO, and from +\$5,000/MWh to -\$100/MWh in ERCOT [8, 20]. Beyond mix, sources of variation include transmission (congestion), fuel prices (natural gas, others), and changes in load and load distribution. Generally in the USA, variation in each dispatch region (Independent System Operator) is independent and unique as cross-grid transmission is sparse. The situation is different in Europe, where the national grids have more crossflow.

The ambitious goals for grid decarbonization (e.g., 100% by 2035 in the United Kingdom, by 2040 in New York, 2045 in California, and by 2050 in the European Union [9, 46, 49]) have been driving the fraction of renewable generation upward rapidly, and in several grids such as California and Germany, renewable generation now regularly surpasses 100%. This produces growing curtailment and negative pricing [1, 12, 33, 52]. Curtailment (overgeneration) occurs when the available power exceeds the load, causing the power to be wasted; it is typically not allowed in the grid and not accounted for as grid generation. Negative pricing is an increasingly frequent phenomenon where a generator is compelled to pay the power grid to accept power. Note that this may be rational because solar and wind generation have nearly zero incremental generation cost and are generally subsidized with *production tax credits* or *feed-in tariffs*. The consequence is that renewable power generation varies both above and below demand as shown in Figure 2.6. This is because solar and wind generally have average generation capacity around 33%, meaning for a 1 GW plant, the average annual generation is $365 \times 24 \times 0.33$ GWh. This means for each GW-year (about 9 TWh) of power needed, you need to build 3 GW of generation capacity. Of course, when the sun is shining and the wind is blowing, this capacity might generate a full 3 GW, well above the average need.

Excess is illustrated in Figure 2.7. Both curtailment and negative pricing are common in all power grids exceeding 30% renewable power fraction. Despite the addition of large quantities of energy storage and advanced load

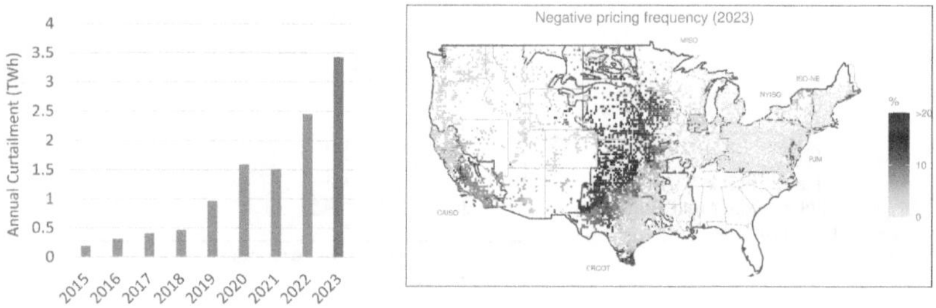

Figure 2.7: Annual renewable curtailment in the CAISO grid (2015-2023), and negative pricing across the USA (2023). In advanced renewable grids, curtailment is gigawatt scale, and negative pricing is occurring with increasing frequency [10, 13, 33].

control (adaptive, shifting, shaping, etc. [7, 22]), the quantity of power subject to these problems will rapidly grow as power grids scale to 50%, 75% and higher levels of renewable penetration. This signature challenge for utilities in the coming decades is the focus of much study [12, 21, 22] to avoid the cost of energy storage [7]. Load flexibility, discussed in Section 3.6.1, is critically important for the technical and economic challenges to solve curtailment and negative pricing in grids with higher levels of grid decarbonization [22, 25, 41, 52, 55, 56].

While curtailment and negative pricing intrigue many people because they are such a novel situation, the problem of undergeneration is more severe. Even with 3 GW of generation, there will be times when there is no sun or wind, this generation might produce well below the 1 GW average demand. This is often viewed as a reliability problem by the power grid and means that load will be shed. That is, some datacenters will not receive their desired power and will have to reduce capacity or even shut down. As a reality, this happened in several regions of Europe in 2022, due to both a widespread drought and a lack of wind [4].

2.3 AI Datacenter Growth Drives Demand for Flexible Load

The rapid increase of AI datacenter load, worldwide, is well documented [5, 6, 11, 43] with projections of datacenter load increase ranging from a 4-10x increase by 2030, reaching 3.5-17% of USA grid total load. The rapid growth threatens grid reliability, incurs large capital costs for new generation and transmission, and can slow or even reverse grid decarbonization [3, 28, 40, 45].

Furthermore, these capital expenses and the large datacenter loads can directly harm other power consumers by raising their prices or attributed carbon. Higher loads threaten grid reliability in the face of increasing renewable generation [30], and to maintain reliability, retaining or even increasing fossil-fuel generation is a capital efficient approach to increasing grid capacity [44][1]. With climate change a global concern, the prospect of significant retardation of power grid decarbonization in the face of challenging climate goals faces strong societal opposition. So AI datacenters especially (but all datacenters) should become grid-integrated, operated as adaptive load in cooperation with the grid.

The rapid growth of datacenters is projected to increase load by 80 GW in the United States from 2023 to 2030. While there is much discussion of nuclear power as a source of carbon-free energy, its timeline is too long to be relevant for at least 10 years (through 2035). So, for the near future, the grid must find a way to add new generation capacity and balance it with the growing datacenter load as illustrated in Figure 2.6. This figure depicts the "incremental" problem of the new constant datacenter load and the new variable generation.

The major approaches available for power grids are (1) overbuild generation and add storage (expensive), or (2) create and exploit flexible load [18, 21, 22]. The latter is most attractive because of its cost-effectiveness and is a major path to enable growth of datacenter load [19].

2.4 Grid-integrated AI Datacenters Are Dynamic and Cooperative

Future datacenters must become grid-integrated, operated as adaptive load in cooperation with the grid. In the past, many power grids have had demand-response programs; these programs focus on peak shaving [18, 38, 47, 51, 59] and as a result they help little with datacenters. The computing services provided abhor even short disruptions and there is little incentive because of the immense profits made with each hour of computing power delivered. Each dollar or euro of electric power is refined into hundreds to thousands of times the value by the datacenter (computing factory for digital services). More radical, less reliable power models such as Zero-carbon Cloud or Bitcoin-style Large Flexible Load [18, 55, 56] that inflict unreliability on computing workloads are considered unsuitable for AI computing workloads by many [53].

[1] As a sign of the desperation for capacity, Japan, UK, and the United States have reversed course and are restarting or even building new nuclear generation to meet the need [36]. However, both technical and economic challenges suggest that significant nuclear capacity appears to be well beyond 2030.

To maintain reliable compute power, and for the grid acceptable loss of load expectation (LOLE) and optimize grid operation, much greater cooperation is needed.

A fundamental reason for power grid variation – in carbon-content and supply – is that typical renewable (wind or solar) has usually a capacity factor of one-third. That is, on average over the year, a 1 MW generator will generate about 0.33 MW-year (18.7 GWh). This means that renewables must be overprovisioned 3-fold to ensure adequate average supply. The overprovisioning produces large swings in generation, and when a larger fraction of generation is producing, the supply can exceed 100% of load (overgeneration). This oversupply produces curtailment and negative pricing, and the complementary shortage produces brownout or carbon emissions from fossil fuel generators. In current market structures, overgeneration produces very low power prices (zero or even negative) and shortages very high prices (e.g., \$5000/MWh in ERCOT). These high short-term variations in power availability, cost, and carbon content represent opportunities for **Dynamic Datacenters**, which can be large adaptive power loads (see Figure 2.8). The emergent problem of

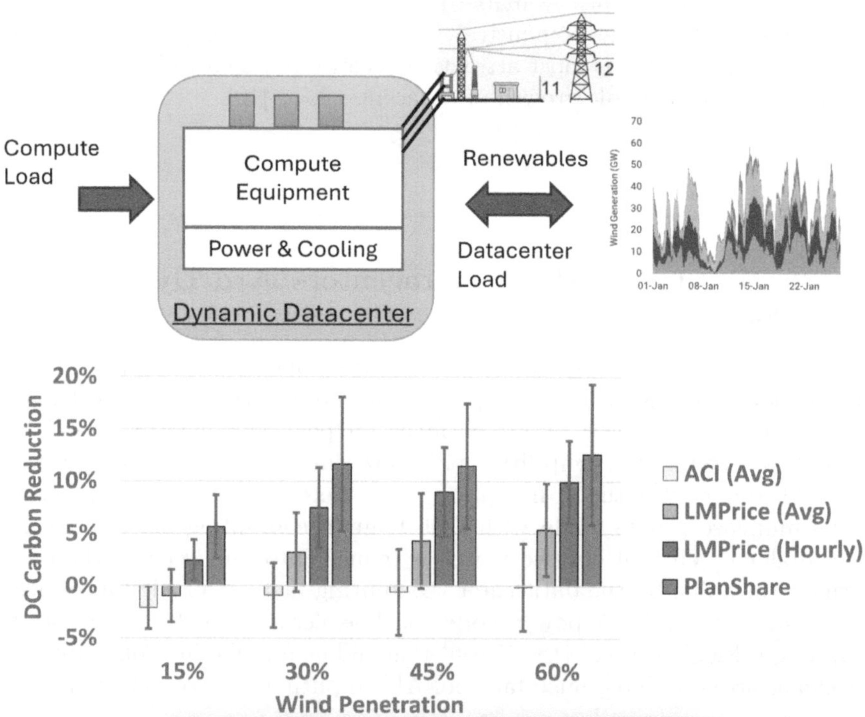

Figure 2.8: A grid-integrated dynamic datacenter that plans and shares its load with the grid, cooperating to reduce carbon emissions (above); potential datacenter carbon emissions benefit from cooperation (below, from [29]).

trough and peak generation is a significant new challenge for power grids [15]; the preferred solution is adaptive loads (e.g., datacenter and other large load flexibility [14, 22, 59]). Thus, as AI datacenters add hundreds of gigawatts load to the world's power grids, they must increasingly become dynamic datacenters integrated as adaptive loads. The alternative is much higher grid costs for both energy storage and "backup" fossil-fuel generation.

As shown in Figure 2.8, a dynamic datacenter can respond to both compute load and grid circumstance (generation, price, carbon, etc.), and it cooperates with the power grid to determine an appropriate power level – now and in the future (plan). This can produce several benefits. First, the availability of large adaptive loads, with some grid control, enables tighter reliability analysis – increasing the amount of load that can be safely supported, and perhaps as important increasing the utilization of generation [24]. Second, the ability to adjust load enables the grid to absorb a larger fraction of renewable generation [27, 29, 55, 56], accelerating grid decarbonization. Third, the natural effect of grid adjustment of datacenter load will be to increase when renewables are plentiful, and decrease when they are scarce. This will reduce the average carbon emissions of the adaptive loads (e.g., datacenters).

As a large and fast growing fraction of the power grid load [17, 39, 50] and to achieve these benefits, AI datacenters must be dynamic and cooperative to help the future power grid grow, maintain reliability, and decarbonize. *Dynamic datacenters* can increase demand when power is plentiful, but they must decrease load when power is in short supply. These datacenters can *cooperate* with the grid, sharing information (load plans) and allowing their power loads to be shaped, enabling the grid to optimize performance (cost, lowest carbon emissions), but also to ensure grid reliability. As shown in Figure 2.8 (right), cooperation can reduce datacenter carbon emissions significantly while also reducing power costs. So these external, fundamental large-scale grid concerns are what give rise to the notion of *variable capacity* computing resources, and the *variable capacity scheduling* problem.

Some active areas of research include:

- **Dynamic Adaptive Loads.** Increasingly flexible datacenter design, operating across a wider range of power levels. Evolution from a static maximum power target (e.g., 500 MW) to design and operation with "dynamic range" of 25% from 375 to 625 MW. These new datacenters would operate in a power-efficient, cost-effective manner across that range [14].

- **Grid Cooperation.** New datacenter to power grid interfaces that enable cooperation with the power grid, negotiating power levels and power-level change rates that are suitable for both (enabling low-cost, low-carbon, reliable grid operation and suitable datacenter efficiency), and do so in an open market-competition setting involving multiple datacenters, and multiple types of adaptive loads from other industries [18, 21, 22, 27, 29, 31, 58]. These efforts will likely provide foresight into variation (predictability) but may ultimately produce a wider range of variation.

- **Shifting for Carbon-reduction.** Beyond grid balancing, many of the dynamic adaptation techniques can also be used to reduce grid carbon emissions, or perhaps more important for computing, the datacenter attributed carbon emissions [27, 29, 42, 58].

2.5 Implications for Scheduling

The implications of dynamic datacenters are significant for scheduling, bringing variability in compute capacity, prediction, and even unexpected change. For additional scheduling criteria such as carbon emissions or power cost, equally complex variation occurs – at times both correlated and uncorrelated with capacity. Such grid-integrated datacenters have a capacity that is determined by a complex mix of dynamic criteria: natural environment (temperature, time of day, season), power markets (competing power load, forecasts, speculative trading, reliability budget), and competitive computing load. Externally, their power is supplied by a fluctuating mix of generation with varying carbon emission properties[2].

The more complex reality of grid-integrated dynamic datacenters gives rise to a number of attributes and dimensions of the scheduling (or resource management problem). We highlight a few of them below.

- **Describing Variable Capacity.** Scheduling formulations often specify resource capacity as a constant, C. With parallel and multidimensional scheduling, this might be elaborated to include multiple resources, and compute and memory attributes $C_{i,s,m}$, where i indicates a node, s a compute rate, and m a memory capacity. However, with dynamic variation in datacenter capacity, there is a choice. A variation in capacity can be mapped into the number of nodes, $range(i)$, or perhaps in some cases into the values of s, the speed of processing or even the memory capacity. In short, all of these attributes become a function of time $C_{i,s,m}(t)$.

  Once the resources vary with time, it is natural to consider their distribution, predictability (accuracy, horizon, confidence) and more. These are all interesting directions of study, as some seem easy to forecast (wind, sun), but are often complicated by competing demand for power. And of course, all of these are ultimately tied to power markets – and such have been shown to be highly volatile and complex to understand.

- **Scheduling for Contingency (risk).** With variation come new dimensions of performance. Jobs or parts of a workload can suffer

[2]It is also true that carbon-accounting for this mix can be based on side contracts (e.g., Power-Purchase Agreements (PPA)). However, we will not discuss such accounting activities, which are generally unrelated to technical realities of power flow.

setbacks – termination due to loss of resources. With unpredictable and unbounded variation, avoiding this completely may be impossible. Hence, scheduling to manage the risks of these adverse outcomes is interesting [37]. Perhaps to address these dimensions, we need new metrics and algorithms. For example, can we bound the termination risk for a job? Order it across jobs? Independent of attributes such as run duration? Number of parallel resources?

- **Beyond Passive Scheduling to Control.** Next, it is worth considering when it is possible to shape variable capacity. While in general, we can think of datacenter capacity being shaped by outside forces, they all coexist in general economic framework of value. Grid reliability and decarbonization can be traded against compute efficiency (scheduling), and in some regimes a scheduler might be able to "demand" resource stability, or even increase. This would require sufficiently high value computing applications. So it is also interesting to consider how a scheduler might control variations of capacity, at a cost relative to the value functions of the computing schedule outcomes. This is another interesting dimension of scheduling study.

  How might this be realized? Power can always be bought at a higher cost ($$ or carbon emissions) in power markets. Or datacenters and compute complexes are increasingly augmented with small amounts of energy storage, giving them some local control.

- **New Metrics.** Finally, we expect that many new metrics characterizing variable capacity resources, perhaps workloads, and the performance of scheduling systems on them are needed. Simple measures such as stretch, slack, and throughput will need to be augmented. One direction is goodput with penalty functions for job loss – of various times for cloud-type and HPC-type workloads. Another direction is the addition of derivatives that mix in carbon intensity and energy efficiency.

2.6 Summary

Datacenters are massive power loads and are colliding with transformative changes in global power grids. Efficient, reliable datacenter operation means that their capacity will increasingly become variable, and that variation is driven by external forces of wind and sun, as well as markets and competition. That reality opens a wide range of new research questions and opportunities; we have touched on a few of them here.

2.7 References

[1] B. Acun, B. Morgan, H. Richardson, N. Steinsultz, and C.-J. Wu. Unlocking the potential of renewable energy through curtailment prediction, 2024. https://arxiv.org/abs/2405.18526 (accessed 12-12-2025).

[2] Aurora exascale computer, June 2024. https://www.anl.gov/aurora (accessed 12-12-2025).

[3] Aventine. What Can We Do About AI's Insatiable Thirst for Energy?, 2024. https://www.aventine.org/energy-consumption-ai-lab-grown-meat (accessed 12-12-2025).

[4] Bloomberg. Google, Oracle Data Centers Knocked Offline by London Heat. *Bloomberg*, 2022. https://www.datacenterknowledge.com/cooling/google-oracle-data-centers-knocked-offline-by-london-heat (accessed 12-12-2025).

[5] Bloomberg. OpenAI Pitched White House on Unprecedented Datacenter Buildout. *Bloomberg*, 2024. https://www.bloomberg.com/news/articles/2024-09-24/openai-pitched-white-house-on-unprecedented-data-center-buildout (accessed 12-12-2025).

[6] Bloomberg. Solar, Gas, and Batteries Race to Power AI Revolution. *bloomberg*, 2024. https://www.bloomberg.com/professional/insights/technology/solar-gas-and-batteries-race-to-power-ai-revolution/ (accessed 12-12-2025).

[7] CAISO. 2023 Special Report on Battery Storage.

[8] CAISO. CAISO Market Data. https://www.caiso.com/todays-outlook/ (accessed 12-12-2025).

[9] Developing renewable energy. https://www.energy.ca.gov/about/core-responsibility-fact-sheets/developing-renewable-energy (accessed 12-12-2025).

[10] A. A. Chien. Characterizing opportunity power in the california independent system operator (caiso) in years 2015-2017. *Energy and Earth Science*, 3(2), December 2020. Also available as University of Chicago, Computer Science TR-2018-07, https://newtraell.cs.uchicago.edu/research/publications/techreports (accessed 12-12-2025).

[11] A. A. Chien. GenAI: Giga$$$, TeraWatt-Hours, and GigaTons of CO2. *Commun. ACM*, 66(8):5, July 2023.

[12] A. A. Chien and L. Lin. As grids reach 100% renewable at peak, growing curtailment of 8 gigawatts looms as a challenge to decarbonization. *SIGENERGY Energy Inform. Rev.*, 4(1):3–10, Feb. 2024.

[13] A. A. Chien, F. Yang, and C. Zhang. Characterizing curtailed and uneconomic renewable power in the mid-continent independent system operator. *AIMS Energy*, 6(2):376–401, December 2018.

[14] A. A. Chien, C. Zhang, and L. Lin. Beyond PUE: Flexible datacenters empowering the cloud to decarbonize. *USENIX Hot Carbon*, 2022.

[15] E3. Investigating a higher renewables portfolio standard in california: Executive summary. Technical report, Report from Energy and Economics, Inc., January 2014.

[16] M. Ebeed, S. Kamel, and F. Jurado. Chapter 7 - optimal power flow using recent optimization techniques. In A. F. Zobaa, S. H. Abdel Aleem, and A. Y. Abdelaziz, editors, *Classical and Recent Aspects of Power System Optimization*, pages 157–183. Academic Press, 2018.

[17] EirGrid Group. Generation capacity statement 2023–2032, 2024. `https://cms.eirgrid.ie/sites/default/files/publications/19035-EirGrid-Generation-Capacity-Statement- Combined-2023-V5-Jan-2024.pdf` (accessed 9-1-2024).

[18] Electric Reliability Council of Texas. Large flexible load task force. `https://www.ercot.com/committees/tac/lfltf` (accessed 1-1-2024).

[19] EPRI Launches Initiative to Enhance Data Center Flexibility and Grid Reliability, October 2024. `https://www.epri.com/about/media-resources/press-release/yimzjv2xnv9cqiztau1zxbedletwyqk1` (accessed 12-12-2025).

[20] Electric Reliability Council of Texas (Texas Power Grid). `http://www.ercot.com` (accessed 12-12-2025).

[21] B. F. Gerke, G. Gallo, S. J. Smith, J. Liu, S. V. Raghavan, P. Schwartz, M. A. Piette, R. Yin, and S. Stensson. The california demand response potential study, phase 3: Final report on the shift resource through 2030, 07/2020 2020.

[22] B. F. Gerke, S. J. Smith, S. Murthy, S. Baik, S. Agarwal, P. Alstone, A. Khandekar, C. Zhang, R. E. Brown, J. Liu, and M. A. Piette. The california demand response potential study, phase 4: Report on shed and shift resources through 2050, 05/2024 2024.

[23] W. E. Gnibga, A. A. Chien, A. Blavette, and A. C. Orgerie. FlexCoolDC: Datacenter Cooling Flexibility for Harmonizing Water, Energy, Carbon, and Cost Trade-offs. In *Proceedings of the 15th ACM International Conference on Future and Sustainable Energy Systems*, e-Energy '24, page 108–122, New York, NY, USA, 2024. Association for Computing Machinery.

[24] J. Horn, Y. Wu, A. Khodabakhsh, E. Nikolova, and E. Pountourakis. The long-term cost of energy generation. In *Proceedings of the Eleventh ACM International Conference on Future Energy Systems*, e-Energy '20, page 74–85, New York, NY, USA, 2020. Association for Computing Machinery.

[25] M. Hoza. Wind and solar curtailments on the rise. `https://insight.factset.com/wind-and-solar-curtailments-on-the-rise` (accessed 12-12-2025).

[26] International Energy Agency (IEA). Electricity 2024, 2024. `https://www.iea.org/reports/electricity-2024` (accessed 12-12-2025).

[27] K. Kim, F. Yang, V. Zavala, and A. A. Chien. Data centers as dispatchable loads to harness stranded power. *IEEE Transactions on Sustainable Energy*, 2016. DOI 10.1109/TSTE.2016.2593607.

[28] S. Kishan and J. Saul. AI Needs So Much Power That Old Coal Plants Are Sticking Around. *Bloomberg*, Jan. 2024. `https://www.bloomberg.com/news/articles/2024-01-25/ai-needs-so-much-power-that-old-coal-plants-are-sticking-around` (accessed 12-12-2025).

[29] L. Lin and A. A. Chien. Adapting datacenter capacity for greener datacenters and grid. In *Proceedings of the 14th ACM International Conference on Future Energy Systems*, e-Energy '23, page 200–213, New York, NY, USA, 2023. Association for Computing Machinery.

[30] L. Lin, R. Wijayawardana, V. Rao, H. Nguyen, E. W. Gnibga, and A. A. Chien. Exploding AI Power Use: an Opportunity to Rethink Grid Planning and Management. In *Proceedings of the 15th ACM International Conference on Future and Sustainable Energy Systems*, e-Energy '24, page 434–441, New York, NY, USA, 2024. Association for Computing Machinery.

[31] J. Lindberg, Y. Abdennadher, J. Chen, B. C. Lesieutre, and L. Roald. A guide to reducing carbon emissions through data center geographical load shifting. In *Proceedings of the Twelfth ACM International Conference on Future Energy Systems*, e-Energy '21, page 430–436, New York, NY, USA, 2021. Association for Computing Machinery.

[32] The Large Unified Modern Infrastructure (LUMI) Supercomputer, November 2024. `https://lumi-supercomputer.eu/lumi_supercomputer/` (accessed 12-12-2025).

[33] D. Millstein, E. O'Shaughnessy, and R. Wiser. Exploring wholesale energy price trends the renewables and wholesale electricity prices (rewep) tool, version 2024. `https://emp.lbl.gov/renewables-and-wholesale-electricity-prices-rewep` (accessed 1-1-2025).

[34] Energy markets 101: Buying and selling energy. https://www.misoenergy.org/meet-miso/market-basics/ (accessed 1-1-2024).

[35] NAE. Greatest engineering accomplishments of the 20th century. https://www.nae.edu/7461/GreatAchievementsandGrandChallenges (accessed 12-12-2025).

[36] I. Penn and K. Weise. Hungry for energy, amazon, google and microsoft turn to nuclear power. *new york times*, October 2024. https://www.nytimes.com/2024/10/16/business/energy-environment/amazon-google-microsoft-nuclear-energy.html (accessed 12-12-2025).

[37] L. Perotin, C. Zhang, R. Wijayawardana, A. Benoit, Y. Robert, and A. Chien. Risk-aware scheduling algorithms for variable capacity resources. In *Proceedings of the SC '23 Workshops of the International Conference on High Performance Computing, Network, Storage, and Analysis*, SC-W '23, page 1306–1315, New York, NY, USA, 2023. Association for Computing Machinery.

[38] PJM. Demand response, 2024. https://www.pjm.com/markets-and-operations/demand-response (accessed 12-12-2025).

[39] PJM. PJM Load Forecast Report, 2024. https://www.pjm.com/-/media/library/reports-notices/load-forecast/2024-load-report.ashx (accessed 12-12-2025).

[40] B. Plumer and N. Popovich. A New Surge in Power Use Is Threatening U.S. Climate Goals, 2024. https://www.nytimes.com/interactive/2024/03/13/climate/electric-power-climate-change.html (accessed 12-12-2025).

[41] M. Putnam and M. Perez. Solar potential analysis report, November 2018. Prepared for Minnesota Department of Commerce and the Minnesota Solar Pathways Project (curtailment more cost-effective than storage), Available from http://mnsolarpathways.org/ (accessed 12-12-2025).

[42] A. Radovanovic, R. Koningstein, I. Schneider, B. Chen, A. Duarte, B. Roy, D. Xiao, M. Haridasan, P. Hung, N. Care, S. Talukdar, E. Mullen, K. Smith, M. Cottman, and W. Cirne. Carbon-aware computing for datacenters. *CoRR*, abs/2106.11750, 2021.

[43] G. Sachs. Generational Growth: AI, data centers and the coming US power demand surge. Technical report, Goldman Sachs, April 2024.

[44] P. Sisson. A.I. Frenzy Complicates Efforts to Keep Power-Hungry Data Sites Green, 2024. https://www.nytimes.com/2024/02/29/business/artificial-intelligence-data-centers-green-power.html (accessed 12-12-2025).

[45] Staff. Asia is weighing data-centre ambitions against sustainability. *Economist*, October 2024.

[46] State renewable portfolio standards and goals. `https://www.ncsl.org/energy/state-renewable-portfolio-standards-and-goals` (accessed 12-12-2025).

[47] Q. Sun, S. Ren, C. Wu, and Z. Li. An online incentive mechanism for emergency demand response in geo-distributed colocation data centers. In *Proceedings of the Seventh International Conference on Future Energy Systems*, e-Energy '16, New York, NY, USA, 2016. Association for Computing Machinery.

[48] The top 500 list (of supercomputers). `https://wwww.top500.org` (accessed 12-12-2025).

[49] Virginia Department of Environmental Quality. Renewable energy, 2024. `https://www.deq.virginia.gov/our-programs/renewable-energy` (accessed 12-12-2025).

[50] Virginia Electric and Power Company. 2023 integrated resource plan, 2023. `https://cdn-dominionenergy-prd-001.azureedge.net/-/media/pdfs/global/company/2023-va-integrated-resource-plan.pdf` (accessed 9-1-2024).

[51] A. Wierman, Z. Liu, I. Liu, and H. Mohsenian-Rad. Opportunities and challenges for data center demand response. In *International Green Computing Conference*, pages 1–10, 2014.

[52] R. H. Wiser, A. D. Mills, J. Seel, T. Levin, and A. Botterud. Impacts of variable renewable energy on bulk power system assets, pricing, and costs. Technical Report LBNL-2001082, Lawrence Berkeley National Laboratory, 11/2017 2017. A link to a webinar recorded on December 13, 2017 can be found at `https://youtu.be/EMrFAklQnPI`.

[53] C.-J. Wu, R. Raghavendra, U. Gupta, B. Acun, N. Ardalani, K. Maeng, G. Chang, F. A. Behram, J. Huang, C. Bai, M. Gschwind, A. Gupta, M. Ott, A. Melnikov, S. Candido, D. Brooks, G. Chauhan, B. Lee, H.-H. S. Lee, B. Akyildiz, M. Balandat, J. Spisak, R. Jain, M. Rabbat, and K. Hazelwood. Sustainable AI: Environmental Implications, Challenges and Opportunities, 2022. `https://arxiv.org/abs/2111.00364`.

[54] Elon Musk's Massive AI data center gets unlocked: xAI gets approved for 150MW of power, enabling all 100,000 GPUs to run concurrently, November 2024. `https://www.tomshardware.com/tech-industry/artificial-intelligence/elon-musks-massive-ai-data-center-gets-unlocked-xai-gets-approved-for-150mw-of-power-enabling-all-100-000-gpus-to-run-concurrently` (accessed 12-12-2025).

[55] F. Yang and A. A. Chien. ZCCloud: Exploring wasted green power for high-performance computing. In *2016 IEEE International Parallel and Distributed Processing Symposium (IPDPS)*, pages 1051–1060. IEEE, 2016.

[56] F. Yang and A. A. Chien. Large-scale and extreme-scale computing with stranded green power: Opportunities and costs. *IEEE Transactions on Parallel and Distributed Systems*, 29(5), December 2017.

[57] C. Zhang, A. G. Kumbhare, I. Manousakis, D. Zhang, P. A. Misra, R. Assis, K. Woolcock, N. Mahalingam, B. Warrier, D. Gauthier, L. Kunnath, S. Solomon, O. Morales, M. Fontoura, and R. Bianchini. Flex: high-availability datacenters with zero reserved power. In *Proceedings of the 48th Annual International Symposium on Computer Architecture*, ISCA '21, page 319–332. IEEE Press, 2021.

[58] W. Zhang, L. A. Roald, A. A. Chien, J. R. Birge, and V. M. Zavala. Flexibility from networks of data centers: A market clearing formulation with virtual links. *Electric Power Systems Research*, 189:106723, 2020.

[59] Y. Zhang, I. C. Paschalidis, and A. K. Coskun. Data center participation in demand response programs with quality-of-service guarantees. In *Proceedings of the Tenth ACM International Conference on Future Energy Systems*, e-Energy '19, page 285–302, New York, NY, USA, 2019. Association for Computing Machinery.

Chapter 3

Scheduling Algorithms and Techniques with Variable Capacity Resources

Anne Benoit, Andrew A. Chien, Florina M. Ciorba, Thomas Hérault, and Yves Robert

This chapter surveys the impact of resource variability onto scheduling algorithms and metrics. First, we survey recent work related to energy-aware scheduling. Next, we discuss characteristics of current and future HPC and datacenters, with a focus on power sources and their variation. Then, we investigate several scheduling techniques that can be deployed for batch schedulers to cope with, or even benefit from, changes in the number of computing resources. The chapter continues with a broad overview of the challenges that face scheduling techniques to account for variability, from new workload types up to combined optimization metrics. We conclude by sketching two case studies, one for applications with variable running times, and one for risk-aware mapping jobs on dynamically changing platforms.

3.1 Introduction

This chapter surveys the impact of resource variability onto scheduling algorithms and metrics, in the context of high-performance computing (HPC) applications and systems. The insights and approaches are equally relevant to cloud computing applications and datacenters, without explicitly differentiating between these environments. In Section 3.2, we start with a brief overview of some recent efforts related to energy-aware and power-aware scheduling. Section 3.3 briefly deals with the current workload characteristics. Then, in Section 3.4, we detail a possible classification of HPC centers depending upon how power is contracted. Section 3.5 provides a detailed description of several scheduling techniques that can be used to extend current batch schedulers, depending upon which type of HPC center is targeted. This concludes our presentation of current and envisioned scheduling algorithms to deal with resource flexibility.

To go beyond this state-of-the-art approach, we study the characteristics of future workloads in Section 3.6 and new optimization metrics in Section 3.7. Future research directions for scheduling are outlined in Section 3.8. Finally, we conclude in Section 3.9.

3.2 Energy-Aware and Power-Aware Scheduling

Several surveys [13, 41, 43, 44] detailed current energy awareness and power management techniques that are regularly applied in HPC centers. In [43], the authors summarize a set of recent efforts to study new techniques by a collection of HPC centers around the world. In particular, those include:

- Experiments to evaluate energy and cost savings by allowing the job scheduler to inform the selection of the power source at RIKEN Japan (the alternative energy sources are limited to the electrical grid and local production of electricity via gas turbines);
- Researchers at Tokyo Tech Japan study the possibility to shut down idle nodes, by coupling the job scheduler decision with the resource manager;
- Certain centers (CEA France, STFC UK, LANL USA, CINECA Italy) evaluate technologies to couple performance metric gathering and job scheduling, in order to report energy usage to the user, or build statistical databases of energy usage to inform future decisions;
- Other centers are experimenting with monitoring and power capping tools to manage power usage and adapt it to variable external conditions (KAUST Saudi Arabia, LRZ Germany).

At the job level, *Dynamic Voltage and Frequency Scaling* (DVFS) is the most used technique to decrease energy consumption. References [5, 21, 22, 25, 26, 31, 58] and Section 3.5.2.1 further discuss this technique. Energy-aware schedulers are introduced in [5, 9, 36, 39]. These schedulers only change the order of execution of the jobs and do not consider varying external conditions, or capacity for the jobs to change their behavior using DVFS.

Recent works have examined the power consumption of HPC jobs on various platforms [52], concluding that:

- Despite total system usage (TSU) exceeding 80%, average power utilization can be up to 30% lower than thermal design power (TDP), which is the power contracted with the electrical provider. Consequently, excess power is available. This motivates resource oversubscription techniques (see Section 3.5).

- Applications exhibit a high variability of power consumption. For example, the median application used is 75% on one system. Few applications justify maximum power (TDP), and 10% of applications use less than 30%.
- Comparing applications run across different platforms, their performance profile can vary significantly across architectures.
- Only a few users account for the highest power-consuming jobs (20% of users consume 80% of the energy), and the power use of applications from a given user varies significantly.

A form of power capping is exploited in [10, 11, 12], by preventing scheduling of high-energy-consuming applications on the platform when needed (does not adapt to changing conditions). Moreover, moldable jobs, which can be executed with various resource allocations, are addressed in [6, 49, 53], and these works show how moldability allows them to adapt the scheduling to energy and power considerations. However, they consider the case of the initial execution of the job only, and with a static estimation of energy and power availability (or cost). Recently, a study of malleability in HPC considers power as a malleable resource, and jobs subject to power capping that adapt their resource usage during runtime as malleable if the system decides when and how resources are reconfigured, or evolvable if the job both decides and applies these decisions [59]. However, the scheduler needs *a priori* information about the impact of reconfiguration to decide whether it is beneficial to the job or not. Such a precise estimation of application performance is hard or costly to obtain, and traditional job scheduling systems are not suitable for malleable jobs.

Altogether, the works discussed in this section are representative of important contributions in the area, but the list is far from exhaustive. However, to our knowledge, no work has yet explored the wide spectrum of techniques that we describe in Section 3.5.

3.3 Workloads

HPC system workloads constitute a diverse mix of many applications that require a variable amount of computing resources – ranging from a single core for a sequential job to capacity applications that use a large portion of the entire system – and for a variable amount of time – ranging from a few minutes in the extreme case of a short or failing run to the longest time allowed on the scheduler queue. Each system workload has its own characteristics, driven by the center mission and its users [4]. Within the same center, systems dedicated to the same mission still exhibit a wide variety of behaviors [56]. A common characteristic of the observed workload is the presence of small jobs (either in time or in number of resources) [3].

Consider the Oak Ridge Leadership Computing Facility (OLCF), and a specific system, Summit, where the mission is different and so is the workload [1]. OLCF has set a target to assign at least 40% of Summit core-hours to support capability applications – defined as applications that use at least 20% of the nodes of the machine. To achieve this goal, a specific program, INCITE [51], was created to allocate computation time to projects on Summit. In 2020, 44% of the core-hours of Summit were running capability applications over all projects, a number that raises close to 50% for the INCITE projects.

The number of small jobs is a significant characteristic of HPC workloads that is exploited by job schedulers to ensure a high TSU. Summit's TSU was above 90% in 2020, and the other systems report similar numbers. Scheduling heuristics are usually simple (typically first come, first served), but to adapt to the changing conditions (jobs that complete before or after their requested wall-clock time), backfilling strategies are used. The efficiency of backfilling increases when a significant part of the workload has a small granularity. Backfilling is crucial in HPC systems, as the wall-clock time accuracy – ratio of the time actually used by the requested wall-clock time – can be low. In [56], it was distributed uniformly between 0 and 1.

Consequently, any job scheduling policy or algorithm must be robust, portable, tuned, and evaluated over a wide range of systems, workloads, and assumptions.

Two key application features are the possibility of taking a checkpoint, and the flexibility in the number of resources enrolled during execution. **Checkpointable** applications can be suspended, saving their execution state, and later resume from that state. In [38], which characterizes a projected load for USA exascale systems, 50% of applications, half the projected load, include checkpointing capabilities. Some applications can be checkpointed at any instant (they are called *preemptible* in the scheduling literature), while some other applications, such as workflow tasks, can only be checkpointed at the end of their execution (saving the output files to be transmitted to successor tasks).

The second key feature is related to the number of resources enrolled during execution [24, 54, 59]:

Rigid. These jobs require a fixed-size allocation, which remains unchanged throughout their lifetime.

Moldable. These parallel jobs execute with a fixed number of processors, but this number is chosen at the time of submission and can vary in range or shape. A typical example is a matrix product for which the user specifies a list of possible processor grid configurations (say, 10×10, 8×12 or 12×8) and the batch scheduler selects one of them. Once a configuration and allocation are chosen, if the resource manager and the batch scheduler are moldable themselves, i.e., can change the allocation for a job, the moldable jobs become more flexible and improve resource utilization compared to rigid jobs and platforms.

Malleable. These parallel jobs execute with a number of processors that can vary over time at the request of the system. A typical example is a parallel job that can survive losing one processor due to a fail-stop error. Another example is a coordinator-worker job whose number of workers can be increased or decreased by the batch scheduler, depending upon resource availability.

Evolvable. These are malleable jobs whose number of resources can be varied during execution at the request of the user. A typical example is a parallel job that, during its execution, will temporarily require more resources (e.g., due to adaptive mesh refinement or adaptive time sub-stepping) to spread out/speed up parts of the simulation, but then it would revert to the initial number of allocated resources.

Moldable jobs have received considerable attention in the scheduling literature because they can easily adapt to the number of available resources, contrary to rigid jobs, which are easier to program and implement. Malleable jobs require the resource manager or system scheduler to request a change in the job's resource allocation, and the jobs to reach by self-reconfiguring accordingly. Many computational kernels in scientific libraries, e.g., Slate [27], are provided as moldable jobs that can be deployed on several 2D processor grid sizes. Job schedulers like load-sharing facilities (LSFs) [32] allow the user to specify a range of resources on which each job can be scheduled. Altogether, moldable jobs represent an excellent trade-off for deploying applications in a flexible but effortless way. However, scheduling variable capacity resources may require more flexibility, such as the one offered by malleable and evolvable jobs.

3.4 Scenarios for HPC Centers

One can envision HPC batch schedulers that are capable of dramatic reductions in energy consumption and that enable exploiting dynamic power supply scenarios. We propose three operating scenarios for HPC centers and introduce six scheduling techniques for each scenario.

TODAY'S HPC CENTERS. Today, most HPC centers have a fixed contractual envelope for power. Power is contracted at fixed cost and carbon emissions (annual average)[1], and datacenter management and resource management is performed without regard to external conditions (see Figure 3.1a). The illustration shows that scheduler decisions affect power use, well below maximum TDP.

[1]For example, LRZ (Leibniz Rechenzentrum) in Munich [15].

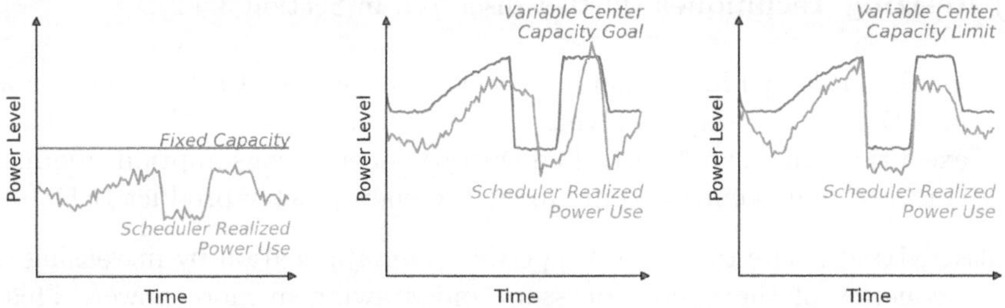

Figure 3.1: Three HPC CENTER scenarios featuring increasing available capacity dynamism and opportunity for growth.

MATCHING HPC CENTERS. These centers recognize the changing external power price and carbon emissions, and target their scheduling to reduce power costs or Scope 2 carbon emissions [15, 35], also known as Operational Expenditures (OPEX) emissions [28] (see Figure 3.1b). Such centers allow us running "power hog" or "hero run" jobs when power is cheap or renewable energy content is high.

FLEXIBLE-POWER HPC CENTERS. The variable power level is enforced on the HPC center by the external environment (local power controller to manage power cost or carbon emissions, power grid for grid stability). The system must operate within the specified variable power envelope (see Figure 3.1c), similar to the idea of a variable power cap [40].

3.5 Algorithmic/Scheduling Techniques

We consider a set of algorithmic/scheduling techniques that can decrease energy consumption and/or carbon emissions, but that also can increase the flexibility of scheduling. We start with a brief presentation of each technique in Section 3.5.1 and discuss its relevance to each scenario for HPC centers. We then describe each technique in further detail in Sections 3.5.2 and 3.5.3.

3.5.1 Outline

We organize the scheduling techniques into two categories.

(A) Resizing Techniques (further discussed in Section 3.5.2):

DVFS. As discussed in Section 3.2, DVFS can be used to slow down an application, decreasing the dynamic energy consumption. However, longer execution time can lead to higher static energy consumption, requiring trade-off management to minimize the energy-delay product (EDP).

TurboMode. The turbo mode speeds up an application by increasing the frequency of the microprocessors and drawing in more power. This is useful, for example, to avoid job interruption if the end of the reservation is nigh. Of course, the turbo mode will consume more power and likely decrease energy efficiency.

ReshapeApplication. Some parallel applications are moldable – they can run with different numbers of processors. For instance, numerical linear algebra kernels can be executed on several 2D processor grid sizes. Typically, user requests specify preferred and permitted configurations. This approach proposes to select an adequate deployment configuration when scheduling a new application and to define incentive policies for users to provide alternative configurations. This approach will also be extended to consider reshaping during execution using SuspendResume.

(B) Reassigning Techniques (further discussed in Section 3.5.3):

SuspendResume. Many parallel applications periodically checkpoint their state, either to be able to restart after a failure or to store partial results before the end of the current allocation. A proactive checkpoint can interrupt an application, free its resources, and allow them to be assigned to another application with higher priority. The checkpointed application can be relaunched with a different configuration (e.g., ReshapeApplication). This combination brings in increased flexibility to the scheduler and creates many opportunities for co-scheduling and application reshaping to meet system and application goals.

ResourceOverSubscription. This approach is both innovative and risky in the context of HPC centers: borrowing from the cloud, one can share computing nodes between two (or more) applications. Sharing nodes and associated performance variation may be preferable when applications fail to use available power. The technique can be combined with ReshapeApplication and SuspendResume: after checkpointing, we might resume execution on a set of shared nodes.

RiskAwareAllocation. The scheduler accounts for the volatility risk levels of different resources. We assign a risk level to each resource (that is, the shut-off priority) and use that information to allocate jobs accordingly. Consider machines ranked from left to right, with safer machines on the left and at-risk machines on the right. At-risk machines will be shut off whenever it is mandatory to reduce power, while safer machines will continue running. A risk-aware scheduler will then assign long

Table 3.1: Suitability of algorithmic/scheduling techniques for current and future HPC centers.

		HPC Centers		
	Technique	Today's	Matching	FlexiblePower
Resize	DVFS	Useful	Useful	Useful
Resize	TurboMode	Inefficient	Useful	Essential
Resize	ReshapeApplication	Useful	Useful	Essential
Reassign	SuspendResume	Useful	Useful	Useful
Reassign	ResourceOverSubscription	Inefficient	Useful	Essential
Reassign	RiskAwareAllocation	Inefficient	Useful	Essential

and/or large jobs to safer machines. See the case study in Section 3.8.2 for more details.

These algorithmic/scheduling techniques often work in combination, and together they open up a large spectrum of new opportunities (and decisions) for the scheduler. For instance, any application resumption can be resized with DVFS, TurboMode, and ReshapeApplication, and/or reassigned with ResourceOverSubscription and RiskAwareAllocation. Moreover, applications that have been resized may require a different scheduling on the allocated resources than before resizing. See the case study in Section 3.8.3 for an approach to automatically select scheduling algorithms. Table 3.1 outlines which method can be applied to which center scenario, showing great synergies.

3.5.2 Resizing applications

3.5.2.1 DVFS

As mentioned in Section 3.2, DVFS is a well-known technique to reduce energy consumption. Processing cores (CPUs, GPUs) can run at several frequencies, whose set is provided by the manufacturer. By default, a nominal frequency is used. Using a lower frequency will decrease the amount of dynamic energy that is consumed but will also increase execution time, thereby increasing the amount of static energy that is consumed. In a standard model, the energy cost of an application running on a single node for a duration d is estimated as the product of power (watt) and duration (seconds), called the power-delay product (PDP) (in joules). The power $P(f_j)$ at frequency f_j is given by

$$P(f_j) = P_{static} + P_{dyn}(f_j) = P_{static} + (P_{indep} + C \times f_j^3),$$

where P_{static} (the static power), P_{indep} (the frequency-independent part of dynamic power), and C (the effective switching capacitance) are system-dependent constants. For a parallel application, one can simply add up the energy cost of all processors, or refine the above simple standard model to include parameters for communications and input/output (I/O) operations.

The performance-energy trade-off achieved by DVFS is at the application level. First, in many cases, slowing down an application and delaying its completion time will not affect the starting time of the next application, because that next application may be waiting for a larger set of resources, so there are no rippling delays in the scheduling. Moreover, the same static energy is spent regardless of whether processors are idle or executing (switching processors off is the only way to spare static energy). At the batch scheduler level, the impact of DVFS can be viewed as follows: applications are rectangles of length, duration, and height, number of processors, and the batch scheduler aims at positioning these rectangles so that they do not overlap, and some metrics, say average response time, is optimized. Then, DVFS is a tool for elongating some rectangles so that (i) some dynamic energy is saved, and (ii) the global positioning of the rectangles remains efficient for the metric (see Application A in Figure 3.2). The trade-off to achieve becomes global to the platform rather than local to each application.

3.5.2.2 TurboMode

Several microprocessors can run in TurboMode, which means that they are overclocked with a frequency higher than the nominal frequency. As such, TurboMode is a particular case of DVFS, where the frequency is deterministically increased rather than dynamically scaled (e.g., Application B in Figure 3.2). Usually, TurboMode comes with a high cost in terms of energy consumption. However, we see its relevance in all scenarios where the number of available resources is going to be reduced due to power capping or increased during a green period, where the carbon intensity is significantly lower than the average carbon intensity for that location [15].

It is better to speed up the execution of a long-running application and terminate its execution before some of its resources become unavailable or the

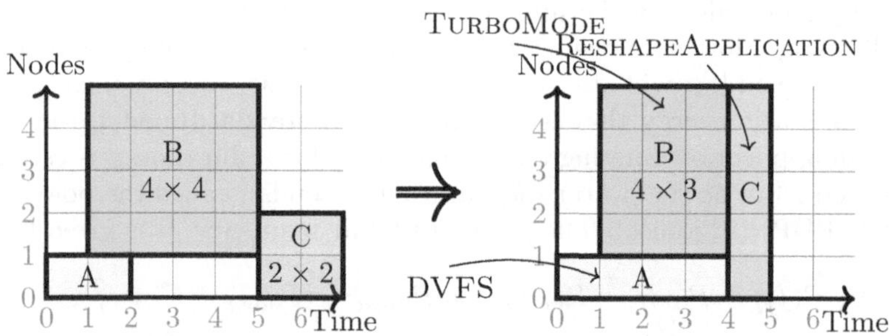

Figure 3.2: Illustration of the resizing techniques for three jobs.
Left: A (2×1), B (4×4), C (2×2), and makespan 7.
Right: A (4×1), B (4×3), C (1×5), and makespan 5.

green period ends. This holds true in particular for capacity applications that require to use most or all computing nodes. Interrupting them because of a resource shortage or ending a green period would lead to a huge waste, because a partial or total re-execution from scratch (depending upon checkpoints having been taken successfully) would (i) consume more energy than using TurboMode to finish up fast, and (ii) considerably delay the response time for the application, put back in the queue, and waiting for enough resources to be free again.

In summary, while it may seem counterintuitive to use TurboMode, which is an energy cruncher, it becomes an essential tool when power variability must be taken into account. However, one must carefully consider the trade-offs, as prolonged execution in TurboMode leads to increased heat dissipation, which in turn leads to a long(er) term (compared to the instantaneous response of DVFS) increase in energy consumption. Running in TurboMode for too long can also lead to a negative effect in which more nodes need to be shut down to account for the increase in energy expense for cooling, even during green periods.

3.5.2.3 ReshapeApplication

As described in Section 3.3, some applications are moldable and can be deployed in a variety of alternative configurations. However, in many HPC centers, nodes need to be allocated hierarchically, following the hierarchy of the machine, to optimize network usage and reduce costly routing operations. In [30], it is shown how the replacement of a node by another in the K-Computer can impact performance, and also that replacing a neighbor node with a distant one could dramatically impact performance.

To allow reshaping applications, we need to take into account the characteristics of the applications as well as the characteristics of the hardware. As this type of reshaping is application-dependent, this decision must be guided by the user. Job schedulers like load-sharing facilities (LSFs) [32] allow the user to specify a range of resources on which the application can be scheduled, but to our knowledge, no current job scheduler deployed in practice allows one to specify complex alternative subsets of constraints. One solution consists of allowing the user to define multiple alternative sets of constraints (a set of constraints typically contains the number of nodes, cores, wall-clock time, number of accelerators, etc.) to run their application, ordered by preference.

Offering alternative sets of constraints can benefit the user in multiple ways: for Today's HPC centers, ReshapeApplication will open new flexibility in scheduling decisions and allow optimizing both resource usage from the platform point of view while reducing or preserving job makespan for the user (see Application C in Figure 3.2). In Flexible-Power HPC centers, the variable power cap will sometimes require a shutdown of computing resources. ReshapeApplication can then be combined with SuspendResume to enable a dynamic reshaping of the

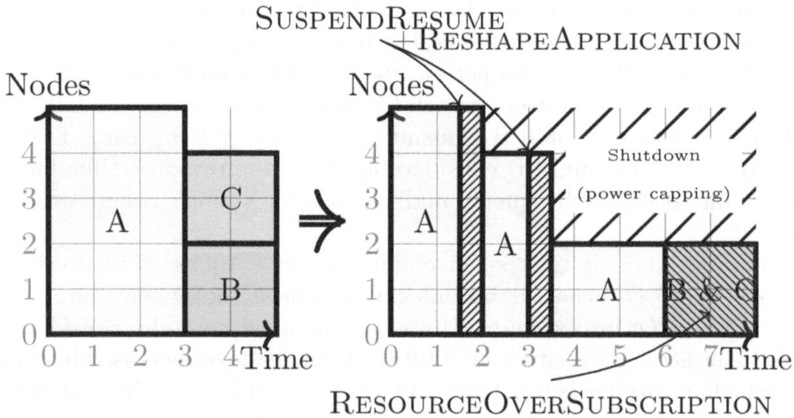

(a) SUSPENDRESUME and RESOURCEOVERSUBSCRIPTION

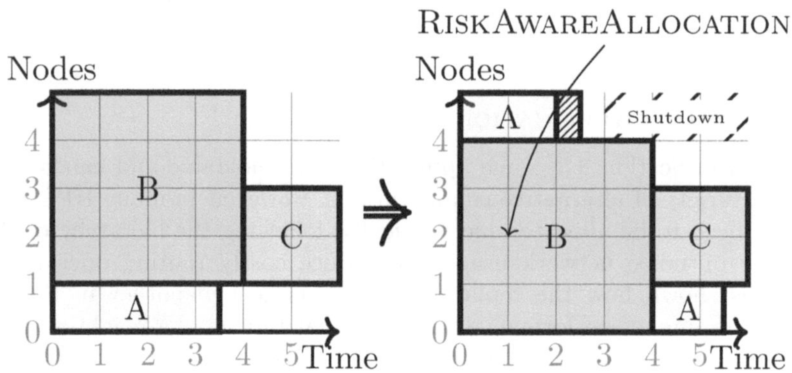

(b) RISKAWAREALLOCATION

Figure 3.3: Illustration of the reassigning techniques. (a) Application A is suspended and resumed twice, while Applications B and C are co-scheduled on the same resources. (b) Application A is suspended and resumed once, while B is allocated to the resources in a risk-aware manner.

remaining available powered on resources. If RESHAPEAPPLICATION is not possible, the job is likely to have to wait until other jobs complete or resources can be integrated back into the platform for it to be resumed. With RESHAPEAPPLICATION, that job has a higher chance of running on the remaining resources, albeit at a slower pace, until additional computing resources can be used again (see Application A in Figure 3.3 for an example). The challenge is very similar for the MATCHING HPC CENTERS, where the energy cost could become too expensive to continue using all the resources.

When reshaping jobs, the user has specified another set of constraints that includes a different wall-clock time. However, in a combined ReshapeApplication/SuspendResume scenario, the application has already spent some time on the machine, which was accounted for by a different wall-clock time assigned to the initial set of constraints. The job scheduler will have to account for the time already spent in relation to the new wall-clock time.

3.5.3 Reassigning jobs to resources

3.5.3.1 SuspendResume

Applications that can produce a checkpoint coordinated among all their processes (or simply checkpoint for brevity) save the progress of their computation at a given time and are able to restart from this intermediate state. Historically, checkpoint/restart has been designed and implemented to provide fault tolerance capabilities, by taking periodic checkpoints and restarting from the last checkpoint in case of node failures or elapsed allocation times. Today, parallel applications use checkpoint/restart for a variety of reasons, ranging from fault tolerance to replay debugging, big data analytics and machine learning, simulation tuning, process migration, and suspend-resume capabilities. The last two features are of particular interest for scheduling variable capacity resources.

One can distinguish three broad categories of **checkpoint/restart**: (1) Application-level checkpointing, (2) Transparent checkpointing, and (3) System-level checkpointing. The former two are applicable for SuspendResume. Many applications of a typical HPC workload feature application-level checkpointing capabilities (e.g., Horovod [57], EAP, LAP, Silverton, VPIC, Dakota, MILC [38]). Application-level checkpointing provides the most diverse uses, as such checkpoints can be used to restart applications in a different configuration (ReshapeApplication), but checkpointing is only possible during specific phases of the application, e.g., between time steps. Transparent checkpointing takes a complete snapshot of the processes' state (memory map, registers, etc.), and uses modern operating systems' capability to reload such images in a portable way. It applies to most HPC applications regardless of their current internal phase, and under some light constraints, can provide a preemptive mechanism to take the checkpoint of a parallel application and restart it. However, processes are restored as they were at the time of checkpoint, so the application cannot be reshaped. It is implemented by libraries like DMTCP [2], which are typically available on HPC systems [65].

Depending on the features of the target application, different checkpointing techniques can be used to provide SuspendResume. The job scheduler will notify the applications to take a checkpoint before a given deadline, after which their execution will be interrupted (see Application A in Figure 3.3(a)). The application will restart from the last checkpoint, so periodic checkpointing should be enabled for suspendable applications, even without considering the risk of node failures, to provide a fallback in case the requested checkpoint does not complete under the deadline.

The time when to suspend an application is obviously at the center of the dynamic job scheduling policies we consider. In TODAY'S HPC CENTERS, SUSPENDRESUME simplifies backfilling by allowing us to schedule long-running applications in parts. For MATCHING HPC CENTERS, it becomes economically viable to suspend high-energy applications in favor of low-energy applications or idling, for periods when the cost of energy is inflated or the carbon intensity is too high. However, suspending (and resuming) an application takes time and thus incurs a cost. For FLEXIBLEPOWER, SUSPENDRESUME becomes necessary, as the nodes may need to be shut down just to meet the variable power cap of such platforms.

In all cases, when checkpointing is used to implement SUSPENDRESUME, interference issues must be taken into account: asking many applications to suspend simultaneously will introduce I/O contention and thus interference between the applications, hence significantly slowing down the checkpointing time and increasing the risk of interruption. This risk must be mitigated by staggering the suspension requests, leading to defining an order of suspension. This is an open problem, but strategies similar to those proposed in [29] can be used.

3.5.3.2 RESOURCEOVERSUBSCRIPTION

Not all applications provide application-level checkpointing, and SUSPENDRESUME can then be implemented externally using transparent checkpointing (see Section 3.5.3.1). Other applications that provide application-level checkpointing might not define a set of constraints that includes a number of nodes compatible with assignments that the scheduling would propose to adapt to the energy or power constraints. In these situations, we propose to study the possibility of oversubscribing computing nodes, by assigning more processes than planned on the subset of nodes that can be allocated to the job, and by allowing different applications to share computing nodes (see Applications B and C in Figure 3.3(a)). RESOURCEOVERSUBSCRIPTION is also known as collocation, co-scheduling, or multitenancy scheduling [23, 34, 60].

This oversubscription can only be done under heavy constraints: the memory requirements of the processes need to fit in the available memory on each node; similar memory constraints also apply if the application uses accelerators. However, RESOURCEOVERSUBSCRIPTION allows for a higher level of flexibility, as has been demonstrated in the cloud, where virtualization is widely used to share computing resources. Consider the drastic case of FLEXIBLEPOWER where too many computing nodes have to be shut down to run more than a single application: either the other applications are delayed, or the resources must be shared!

Opening the possibility of node sharing to other applications will have an impact on their respective performance, and this impact must be carefully evaluated to scale up the time requested by both applications. However, opening this possibility also allows the scheduler to provide progress for multiple applications, implementing a lower maximum stretch (see Section 3.7 for more details on the *stretch* metric).

This node-sharing policy can only be implemented for specific loads: time-dependent jobs (typically jobs that run performance evaluations) will be submitted to queues that do not allow ResourceOverSubscription. Thus, the scheduler will need to find trade-offs between hard and soft scheduling constraints. In addition to reducing the average maximum stretch, thus providing shorter waiting times, the scheduling policy will provide additional incentives for users to promote the submission of jobs that allow ResourceOverSubscription. Such incentives can typically take the form of reduced consumption of quotas [15, 35].

3.5.3.3 RiskAwareAllocation

This technique targets HPC centers that can adapt to variations in power cost and source (carbon intensity). RiskAwareAllocation is key to the efficient operation of Flexible-Power HPC centers but can also be used, up to a limited extent, for Matching HPC centers. When power decreases, the main question is: which machines to power off, and hence which jobs to interrupt (suspend) and reschedule. The rescheduling of interrupted jobs raises three subsequent questions: (1) Do we have to take a proactive checkpoint before the interruption? (2) Which priority should be given to each interrupted job? (3) Which geometry and which nodes are to be assumed for job re-execution?

The answer to the main question (which jobs to interrupt) depends on the answer to the subsequent three questions, because the answer to each of them will have an impact on the optimization criteria, be they performance-related, cost-related, energy-related or combinations thereof (see Section 3.7). The answer to subsequent question 1 (are we notified to checkpoint ahead of a power change?) depends upon the accuracy of the available tool for power change prediction; both its precision (related to false positives) and its recall (related to false negatives) matter. Resource variation in power obeys specific parameters whose evolution is dictated by a mix of technical availability, weather, and economic conditions; so relying on an accurate external predictor may be too optimistic. Of course, we know that more green power will be available on sunny and/or windy days, but going beyond coarse predictions is challenging.

As for questions 2 and 3, everything is to be invented! We envision a risk-aware batch scheduler that is provided with information on the liveness probability of each node, i.e., the interruption risk incurred to use that node, which may itself vary over time. In addition, the scheduler can decide which node to shut down or restart if the power goes up (see Applications A and B in Figure 3.3). Nodes can be ordered according to nonincreasing risk, say from left to right, with safe nodes on the left and risky nodes on the right.

The scheduler would (i) shut down nodes starting from the right, and (ii) assign priority jobs, such as large jobs, to nodes on the left. Of course, the global load of the platform must remain balanced. This calls for sophisticated algorithms that go far beyond the first-fit decisions (with variants) that are used by current batch schedulers.

3.6 Future Research: Workloads

To go beyond state-of-the-art algorithmic/scheduling techniques, we discuss here important characteristics expected of future workloads executing on systems with variable capacity resources.

3.6.1 Flexibility

Current workloads comprise a diverse mix of applications, offering increased flexibility for scheduling. It seems very important that future workloads become, if possible, even more flexible, to allow adapting to the capacity of the platform and effectively utilizing available resources [45, 55]. There are several facets to flexibility for future workloads, which we detail below.

(1) There should be flexible start dates and flexible use of resources for any given workload. Rather than having a rigid requirement to start immediately, workloads should be designed to be executed anytime before a flexible deadline. Unless it is a time-sensitive task (like a weather forecast), we should run any workload whenever enough resources become available, rather than discarding it if it cannot run immediately. Furthermore, when requesting resources for a job, the number of cores needed is often fixed. However, a more dynamic approach involves placing the job in a queue, and as resources become available, the system chooses the allocation based on defined priorities and could further adapt. This allows for better resource management and efficient execution. We should be able to give incentives to users so that they propose flexible workloads.

(2) Flexibility can also be achieved in terms of accuracy. This involves the concept of approximate computing, where less-precise results are acceptable within certain bounds. By defining a function that improves over iterations, the number of iterations can be determined to achieve an acceptable level of precision. This approach allows for trade-offs between computation time and precision.

(3) The primary goal should be to reduce the carbon footprint generated by computing activities. Currently, we focus solely on the ability to run applications, but we should also look at the environmental impact. By shifting the focus to carbon reduction, greater attention can be paid to mitigating the ecological consequences of computing. To do so, we must assess the cost implications of different approaches and resource allocation strategies. This

involves assessing how expensive certain operations or computing tasks can be and incorporating a cost function to make informed decisions based on cost factors.

(4) Service level agreements (SLAs): Cloud platforms are commonly using SLAs, and perhaps HPC platforms can adopt and adapt such ideas and solutions from the Cloud, as there is no SLA in most current systems.

(5) Locality of data: Current (and future) workloads are (and will be) handling large amounts of data. Due to data governance, gravity, and privacy, moving data is likely to be difficult and/or costly. However, flexible workloads should be able to be moved together with their data if this allows for a more efficient execution, reducing the carbon footprint, as highlighted above.

(6) Workflows are a good way to represent flexibility. If the application is made of different parts, the parts can be moved more easily to different locations so that they can take advantage of variable capacity at the platform level. There are different ways of defining workflows in many languages, for example, Taverna, Nextflow, SWIRL, Condor [17, 19, 61].

3.6.2 Research areas and questions

Against the background of increased flexibility of future workload, we frame here research areas and questions to address in the future.

What will/should future workloads look like? Should future workloads have full awareness of their execution requirements? How should uncertainties be taken into account in workload design? What types of jobs are best suited for flexibility? How should workload specifications be defined to incorporate flexibility? Is it advantageous to advocate for non-urgency whenever possible? Is there a need for flexibility in terms of precision allowing for approximate computing?

In addition to these questions, it is crucial to consider different categories of workload urgency and timing, such as mandatory and instant, mandatory and non-instant, and non-mandatory and (non)instant? How should workload representation be approached to capture these nuances effectively? Should we consider malleable jobs and suspendable jobs as much as possible?

How to promote the (necessary) change to flexible HPC applications? How can we motivate users to design applications with as much flexibility as possible? As incentives, pricing policies can be considered, as well as educating the user about the environmental impact and how flexibility can reduce the carbon footprint. For the sake of transparency, it should be clearly stated what happens to jobs with lower priority. In addition, the user SLAs should be respected. An important role will be played by educating developers (users) that flexible HPC applications will also be completed faster. Hence, there is a natural incentive to frame HPC applications as flexible as possible.

How to motivate flexibility and guarantee that there will not be too much loss in performance? Can we identify stragglers and migrate jobs as needed, in order to reach trade-offs between performance and energy consumption? If a task takes too long to complete, we would like to be able to interrupt it if it is a mandatory task and eventually migrate it to a processor where it could complete in a more timely way.

Which kind of flexibility is interesting for workloads? When considering valuable flexibilities for workloads, important aspects include workload grouping for cost-effective migration, running at lower precision with the ability to do additional computations later for improving accuracy, and incorporating affordable interruptions through checkpoints and non-volatile memory. These flexibilities optimize resource utilization, minimize costs, and improve the efficiency and reliability of workloads.

Which kind of flexibility is interesting for infrastructures and hardware? Evaluating the effectiveness of flexible workloads involves various approaches. Conducting experiments on real systems is one method, although it can be costly and time-consuming. Simulations using tools like SimGrid [14] offer a more efficient alternative, providing a simulated environment to evaluate workload performance. In addition, in-house simulators can be developed to model the behavior of real systems using specific assumptions for a given approach.

To approximate the execution time of flexible workloads, the ElastiSim simulator [50] can be used, as it is the first simulator specifically designed to support malleability, allowing the evaluation of adaptability and performance of the workload in different scenarios. By employing a combination of real system experiments, simulations using tools like SimGrid, and the use of specialized simulators like ElastiSim, comprehensive evaluations of flexible workloads can be conducted efficiently and effectively.

3.7 Future Research: Metrics

Part of the difficulty of scheduling on variable resources comes from the wide spectrum of scheduling techniques that can be deployed to guide and improve resource management decisions taken by the batch scheduler. Another part of the difficulty comes from the many conflicting optimization objectives or metrics. We categorize below the need for new metrics to evaluate the performance of modern systems with variable capacity resources.

Performance.

- *Platform-oriented objectives* include platform throughput or *goodput* and defined as the fraction of time resources are involved in productively executing jobs (not idle or computing jobs that will be interrupted).

- *User-oriented objectives* include minimization of response time, also called *flow* and defined as the time spent by the job in the system from submission to full (successful) completion, and include apart from pure execution time, the time spent waiting in the queue, time to suspend/restart/checkpoint the job. The flow objective comes in many flavors: maximum, average, and weighted by application size. For fairness, the flow is often scaled by the base execution time (without interruption nor checkpoints), leading to *stretch*, with the same flavors (maximum, average, weighted by application size).

Cost. Cost-related objectives involve energy consumption, the volume of carbon emissions, their intensity, and the amount of power used. More generally, the total cost of ownership (TCO) includes both capital (HPC center building, embedded carbon) and operational (cooling, power, IT operation, carbon emissions) expenditures. Depending on the power source (the greener, the better), these costs have different weights. An extreme example is the use of stranded power, which can be free [62].

In addition to traditional metrics of efficiency and responsiveness, metrics must now take into account environmental issues (also known as non-performance attributes) such as e-waste, datacenter water use, power, energy consumption and carbon emissions, as well as network congestion and system heterogeneity, discussed in Chapter 2. Capping and reducing brown energy consumption are becoming increasingly important, which involves lowering the power level bought on the fixed annual contract and intermittently increasing the power level using green energy and daily contracts. Furthermore, there is a growing need to shift from "fast" to "green" computing, which emphasizes energy efficiency and sustainability in addition to performance.

3.8 Future Research: Scheduling

Parallel and distributed computing environments used to consist of stable capacity resources, with homogeneous or slightly heterogeneous architectures, and executing jobs with equal priority. The approaches used to model and evaluate performance were agnostic to environmental issues and included platform-oriented metrics such as utilization or goodput, as well as user-oriented metrics like response time and stretch. Goodput is also a variant of platform utilization, where only work executed for successful jobs is accounted for.

With the advent of more complex and diverse parallel and distributed computing environments with resource variability, traditional scheduling models and metric approaches are no longer sufficient. In the following, we highlight important research areas and open questions to be addressed in the future.

1. Modern computing environments require the development of new models for resource variability. These models must consider both homogeneous and heterogeneous resources, as well as the availability of machines, which can vary over time depending on factors such as user demand or energy availability in a geographic location (for example, 'follow the sun') [18, 55]. Furthermore, the concept of system malleability has emerged, which refers to both (i) the ability to adapt to changing resource availability by dynamically adjusting resource allocation and scheduling, and (ii) the ability to increase or decrease the number of required resources in response to the requirements of flexible workloads [59].

2. Modern computing environments also require new models for job classification [33] (while also taking into account privacy aspects). These models must consider "not instant" computing, which allows for delays in starting and completing jobs, as well as the option of approximate computing [46] by specifying mandatory or non-mandatory parts of a job to trade off the accuracy or resolution of the results with the completion of the job. These models must also provide flexibility in iteratively refining these non-/mandatory parts of a job and account for data and work locality and estimation of the data volumes involved.

3.8.1 Research areas and questions

We highlight in the following important research areas and opportunities for moving the field forward toward scheduling of variable capacity resources.

Opportunities for different job types. The new resource and job classification models and evaluation metrics for emerging computing environments with variable capacity resources require the development of new job scheduling algorithms.

From the point of view of the batch job scheduler (for instance, SLURM [63]), the traditional rigid jobs will be complemented by moldable, malleable and evolvable jobs, for which the allocated resources will vary over time, either triggered by the system at submission time (as is the case for moldable jobs), by the system during execution (as is the case for malleable jobs), or by the application during execution (as is the case for evolvable jobs).

For rigid jobs, there are significant opportunities for scheduling, but allocated machines no longer support equal capacity computing. The scheduler must thus be risk-aware, and consider machines with high interruption risks when backfilling jobs of lower priority or higher tolerance to risks.

For all jobs, placement is increasingly important to maintain locality, which may significantly reduce compute and communication energy costs and delays. Based on the constraints and fluctuations of the power grid, the scheduler will have to decide which nodes will be on or off and for how long. The scheduler may also overdrive or use Dynamic Voltage and Frequency Scaling (DVFS) on allocated resources to finish jobs sooner, during green periods, to use stranded power, or to increase goodput. Oversubscription is an increasingly common approach, initially used in smaller compute centers but recently present in top-tier supercomputers (e.g., LUMI [42]). In constant resource capacity scenarios, job preemption and suspension remain sensible approaches, with checkpointing as the de facto mechanism, which requires non-volatile memory.

In the case of moldable jobs, the system decides the number of resources to allocate at job submission time, based on user-provided possible alternative configurations. Therefore, the scheduler must also decide the shape of jobs (degree of parallelism, job length) to allocate the resources, and the shape can vary across different executions. With queues that support preemption and suspension of running jobs, it is important that system schedulers support resumption of a previously suspended job under a different shape than the initial shape.

Malleable jobs, for which the system decides at runtime how many and which resources to allocate, require the same considerations as moldable jobs (above). However, reshaping of jobs may even take place during execution to avoid suspending them (via checkpointing) and to allow for on-the-fly redistribution of the memory footprint. The goal is to shave the load now and shift it later in time. Non-volatile memory will be required to support reliable suspension, checkpointing, and reshaping.

Finally, evolvable jobs, which allow the user to decide their size during execution, are important to be supported by batch schedulers offering more or fewer resources for adaptive resizing under the constraints discussed above.

New scheduling algorithms. There are several approaches to consider for designing new job scheduling algorithms on parallel and distributed computing systems with variable capacity.

First, it is crucial to have an exchange of information between jobs and the job scheduler as a prerequisite. Recent efforts show both the challenges and promise of this approach [20, 47]. This exchange will provide transparency and visibility about what the job scheduler will do with non-instant jobs (which accept execution delays) and whether it respects the users' Service-Level Objectives (SLOs). Establishing effective communication and coordination between jobs and job schedulers requires defining compelling incentives and rewards. Pricing models that incentivize later deadlines could also be explored.

Another important aspect is supporting the batch scheduler with implementation of job checkpointing and reshaping. The batch scheduler should also be risk-aware regarding available resources for allocation and their variable capacity. It is necessary to incorporate a global perspective of jobs' new parameters (as described above in opportunities for different job types), as well as their impact on resource usage. Lastly, being aware of the current state of the platforms' capacity is crucial for effective scheduling on variable capacity resources.

Probabilistic models for resource variability. Models of resource availability have a great impact on scheduling algorithms. There are several variables involved:

- The available power sources will vary with time. This leads to partitioning each day or week into time intervals, during which power will remain of constant intensity and of same nature or mix. Are these intervals known deterministically? If so, how much in advance? If not, can these intervals be modeled to allow for their prediction?
- From one time interval to the next, power resources will change, inducing the datacenter to shut down some resources (power decrease) or enroll some new ones (power increase). To what extent will power change from one interval to the next? And again, is the amount of variation known in advance? If not, can it be predicted accurately?

The range of scenarios has a wide spectrum, from situations where the scheduler is either fully informed of the rate and amount of resource variations at one extreme or completely agnostic of those rates and amounts at the other extreme.

3.8.2 Case study: Risk-aware scheduling algorithms

Some preliminary scheduling algorithms have been designed in a setting where the available power varies over time in an unpredictable way and no checkpointing is allowed, hence requiring to interrupt some jobs when some machines must be shut down due to a decrease in power supply.

In this setting, we have designed novel risk-aware scheduling and mapping algorithms that assign the right job to the right machine, optimizing for the system's goodput, which is defined as the fraction of time devoted to effective computations (i.e., re-execution does not count) [7, 64]. These algorithms employ a variety of techniques to mitigate the impact of resource variation, including maintaining a risk index per machine, mapping longer jobs to safer machines, maintaining local queues at machines, re-executing interrupted jobs on new machines, and redistributing pending jobs as resource capacity increases.

Simulations were performed to evaluate the performance of the risk-aware scheduling heuristics, using a workload trace from Google's Borg system and three synthetic traces. They showed significant gains over first-fit algorithms with an increase of up to 10% goodput, with no loss in complementary metrics, such as the maximum and average stretch [8]. From this case study, it becomes clear that standard first-fit algorithms are insufficient for future variable capacity environments and require re-design in order to maintain the expected level of performance.

This case study with several hard constraints (no possibility to checkpoint, no knowledge about variation in power) leads the way to plenty of future research directions, including the exploration of different workloads, different job execution models (e.g., migration or deferral), different variation models, and malleable workloads that were discussed in Section 3.3.

3.8.3 Case study: Scheduling algorithm selection for applications with variable runtime characteristics

As modern scientific and data science applications grow in complexity, their computational and memory requirements often vary dynamically during runtime, challenging traditional scheduling methods, especially in HPC environments, where task and system heterogeneity coexist.

The fluctuating workload characteristics of HPC and data science applications include load imbalance and shifting compute or memory demands from branching code, data-dependent execution paths, or application resizing (see Section 3.5.2). These dynamic behaviors require an adaptive approach to scheduling that is capable of responding to runtime variability.

Reinforcement Learning (RL)-based scheduling algorithm selection methods have recently been studied for OpenMP applications [37]. Unlike traditional expert-based learning, RL methods adapt dynamically by learning from runtime performance feedback, making them particularly suitable for applications with unpredictable or evolving characteristics. They can effectively identify optimal scheduling algorithms for varying application behaviors, at the cost of exploration during the learning phase.

Case Study. To mitigate the exploration overhead of RL-based selection methods, a hybrid approach [16] combines expert-derived selection methods [48] with RL-based learning [37]. The use of an *expChunk* parameter, derived from expert knowledge [48], significantly reduces selection overhead and improves scheduling efficiency by optimizing chunk sizes for dynamic scheduling algorithms.

Insights. The performance of RL- and expert-based scheduling was evaluated on six diverse applications, each with distinct runtime characteristics, across three HPC systems [16]. Applications with increasing workload imbalance or with varying time-step behaviors, show the need for dynamic scheduling algorithm selection methods. Key observations include: (1) RL-based algorithm selection methods adapt to runtime changes effectively, selecting high-performing scheduling algorithms after a learning phase. (2) The initial learning phase can introduce significant performance degradation, especially for short-running applications. (3) Combining RL methods with expert knowledge (e.g., *expChunk*) improves both runtime adaptability and efficiency.

Broader Implications. This case study emphasizes the critical need for a dynamic approach to scheduling in environments where application characteristics vary significantly during runtime, in complement to resource variations, such as those described in the previous case study (see Section 3.8.2). While expert-based scheduling algorithm selection methods are efficient for well-understood scenarios, RL-based approaches provide a robust solution to adapt to unforeseen workload patterns. The hybridization of these methods offers a promising direction for future research, particularly for applications operating in environments with dynamically fluctuating or constrained resources.

Addressing both dynamic workloads (Section 3.8.3) and runtime resource variability (Section 3.8.2) contributes to the broader discussion of scheduling variable capacity resources, offering promising research directions for practical strategies that optimize application and system performance.

3.9 Conclusion and Future Working Directions

The advent of more complex and diverse computing workloads, resources, and environments demands a reevaluation of traditional workloads, scheduling models, and metrics, which no longer serve the affected communities. This chapter explored the intricate relationship between resource variability and scheduling in HPC systems, highlighting both current methodologies and potential advancements. It discussed future workload characteristics, novel optimization metrics, and emerging research directions, laying the groundwork for innovative scheduling strategies that can adapt to the evolving demands of computing environments.

We conclude with open questions for future work, related to workloads and their models, computing platforms, scheduling objectives, and approaches to support scheduling on variable capacity resources.

Models. How do we expose, express, exchange and exploit all this (new) information about the workloads (shape flexibility and service-level objectives) and the platforms (variable capacity) to a scheduler?

Workloads. How do we express and support the suspendability, evolvability, moldability, and malleability of jobs? How to combine these considerations into combinations of non-mandatory and non-instant computing job requirements?

Platforms. Under the premise that capacity resources exhibit high variability, how can arriving jobs be classified into rigid, moldable, malleable, and evolvable? How can we predict the flexibility of a workload in terms of job shaping? How to model and predict system malleability?

Uncertainty. Both workloads and platforms include high degrees of uncertainty. A workload with uncertain requirements can lead to overestimating or underestimating resource requirements. Platforms with high capacity variations may lead to interrupting or reshaping applications. How to account for uncertainty, on a global level, in scheduling and resource management algorithms?

Objectives. How do we expose and exchange information between jobs and schedulers to enable seamless cooperation? What is the interplay between the user-facing incentives and the job scheduling objectives (which include delays, carbon savings, migration, etc.)? Do we need pricing models and negotiations to model and study this interplay? Do we still want a single-objective optimization function? Or should we consider multiple objectives at a time but optimize for one at a time: performance capping and energy minimization (not always *-capping and performance maximization).

Algorithms. What are the scheduling techniques that allow us to move from single platforms to distributed platforms, in order to exploit workload flexibility and variable resource capacity? Are online algorithm selection approaches a viable solution for adaptive job scheduling on variable capacity resources?

Approaches. How can the new job types be supported by the new job schedulers in practice (rigid+suspendable, moldable+suspendable, malleable+suspendable, evolvable+suspendable)? Do we need a time-quantum approach to support preemption in job scheduling (à la Operating System scheduling)?

Finally, a comprehensive survey and/or a systematic review of existing work on variable capacity resources is urgently needed to consolidate knowledge, identify gaps and guide the next wave of research in this domain.

3.10 References

[1] S. Abraham et al. US department of energy, office of science high performance computing facility operational assessment 2020. Technical Report ORNL/SPR-2021/1950, Oak Ridge Leadership Computing Facility, Oak Ridge National Laboratory, 2021.

[2] J. Ansel, K. Arya, and G. Cooperman. DMTCP: Transparent checkpointing for cluster computations and the desktop. In *2009 IEEE International Symposium on Parallel & Distributed Processing*, pages 1–12. IEEE, 2009.

[3] F. Antici, M. Seyedkazemi Ardebili, A. Bartolini, and Z. Kiziltan. PM100: A job power consumption dataset of a large-scale production HPC system. In *Proceedings of the SC '23 Workshops of The International Conference on High Performance Computing, Network, Storage, and Analysis*. ACM, 2023.

[4] K. Antypas, B. Austin, T. Butler, R. Gerber, C. Whitney, N. Wright, W. Yang, and Z. Zhao. NERSC workload analysis on hopper. Technical Report 6804E, Lawrence Berkeley National Laboratory, 2014.

[5] A. Auweter, A. Bode, M. Brehm, L. Brochard, N. Hammer, H. Huber, R. Panda, F. Thomas, and T. Wilde. A case study of energy aware scheduling on Supermuc. In *International Supercomputing Conference*, pages 394–409. Springer, 2014.

[6] P. E. Bailey, A. Marathe, D. K. Lowenthal, B. Rountree, and M. Schulz. Finding the limits of power-constrained application performance. In *Proc. of the Int. Conf. for High Performance Computing, Networking, Storage and Analysis*, 2015.

[7] J. Basney and M. Livny. Improving Goodput by Coscheduling CPU and Network Capacity. *IJHPCA*, 13(3):220–230, 1999.

[8] M. A. Bender, S. Muthukrishnan, and R. Rajaraman. Improved algorithms for stretch scheduling. In *Proc. of the 13th Annual ACM-SIAM Symp. on Discrete Algorithms*, SODA'02, page 762–771. SIAM, 2002.

[9] A. A. Bhattacharya, D. Culler, A. Kansal, S. Govindan, and S. Sankar. The need for speed and stability in data center power capping. *Sustainable Computing: Informatics and Systems*, 3(3):183–193, 2013.

[10] A. Borghesi, A. Bartolini, M. Lombardi, M. Milano, and L. Benini. Predictive modeling for job power consumption in HPC systems. In *International conference on high performance computing*, pages 181–199. Springer, 2016.

[11] A. Borghesi, F. Collina, M. Lombardi, M. Milano, and L. Benini. Power capping in high performance computing systems. In *International Conference on Principles and Practice of Constraint Programming*, pages 524–540. Springer, 2015.

[12] A. Borghesi, C. Conficoni, M. Lombardi, and A. Bartolini. Ms3: A mediterranean-stile job scheduler for supercomputers-do less when it's too hot! In *2015 International Conference on High Performance Computing & Simulation (HPCS)*, pages 88–95. IEEE, 2015.

[13] K. W. Cameron, R. Ge, and X. Feng. High-performance, power-aware distributed computing for scientific applications. *Computer*, 38(11):40–47, 2005.

[14] H. Casanova, A. Giersch, A. Legrand, M. Quinson, and F. Suter. Versatile, scalable, and accurate simulation of distributed applications and platforms. *J. Parallel and Distributed Computing*, 74(10):2899–2917, 2014.

[15] M. Chadha, E. Arima, A. Raoofy, M. Gerndt, and M. Schulz. Sustainability in HPC: Vision and Opportunities. In *Workshops of The International Conference on High Performance Computing, Network, Storage, and Analysis (SC-W 2023)*. ACM, 2023.

[16] F. M. Ciorba, A. Mohammed, J. H. M. Korndörfer, and A. Eleliemy. Automated Scheduling Algorithm Selection in OpenMP. In *2023 22nd International Symposium on Parallel and Distributed Computing (ISPDC)*, pages 106–109, 2023.

[17] R. F. da Silva et al. Workflows Community Summit 2024: Future Trends and Challenges in Scientific Workflows. Technical report, ORNL, Oct. 2024.

[18] M. D'Amico and J. C. Gonzalez. Energy hardware and workload aware job scheduling towards interconnected HPC environments. *IEEE Transactions on Parallel and Distributed Systems*, 2021.

[19] P. Di Tommaso, M. Chatzou, E. W. Floden, P. P. Barja, E. Palumbo, and C. Notredame. Nextflow enables reproducible computational workflows. *Nature Biotechnology*, 35(4):316–319, Apr. 2017.

[20] A. Eleliemy and F. M. Ciorba. A Resourceful Coordination Approach for Multilevel Scheduling. In *High Performance Computing and Simulation (HPCS) 2020*, 2021.

[21] M. Etinski, J. Corbalan, J. Labarta, and M. Valero. Optimizing job performance under a given power constraint in HPC centers. In *International Conference on Green Computing*, pages 257–267. IEEE, 2010.

[22] M. Etinski, J. Corbalan, J. Labarta, and M. Valero. Parallel job scheduling for power constrained HPC systems. *Parallel Computing*, 38(12):615–630, 2012.

[23] Y. Fan, Z. Lan, P. Rich, W. Allcock, and M. E. Papka. Hybrid Workload Scheduling on HPC Systems. In *2022 IEEE International Parallel and Distributed Processing Symposium (IPDPS)*, pages 470–480, May 2022.

[24] D. G. Feitelson and L. Rudolph. Toward convergence in job schedulers for parallel supercomputers. In *Job Scheduling Strategies for Parallel Processing*, pages 1–26, Berlin, Heidelberg, 1996. Springer Berlin Heidelberg.

[25] F. Fraternali, A. Bartolini, C. Cavazzoni, and L. Benini. Quantifying the impact of variability and heterogeneity on the energy efficiency for a next-generation ultra-green supercomputer. *IEEE Transactions on Parallel and Distributed Systems*, 29(7):1575–1588, 2017.

[26] V. W. Freeh, D. K. Lowenthal, F. Pan, N. Kappiah, R. Springer, B. L. Rountree, and M. E. Femal. Analyzing the energy-time trade-off in high-performance computing applications. *IEEE Transactions on Parallel and Distributed Systems*, 18(6):835–848, 2007.

[27] M. Gates, A. Abdelfattah, K. Akbudak, M. A. Farhan, R. Alomairy, D. Bielich, T. Burgess, S. Cayrols, N. Lindquist, D. Sukkari, and A. YarKhan. Evolution of the SLATE linear algebra library. *Int. J. of High Performance Computing Applications*, 39(1):3–17, 2025.

[28] U. Gupta, Y. G. Kim, S. Lee, J. Tse, H.-H. S. Lee, G.-Y. Wei, D. Brooks, and C.-J. Wu. Chasing carbon: The elusive environmental footprint of computing. In *2021 IEEE International Symposium on High-Performance Computer Architecture (HPCA)*, pages 854–867. IEEE, 2021.

[29] T. Herault, Y. Robert, A. Bouteiller, D. C. Arnold, K. B. Ferreira, G. Bosilca, and J. J. Dongarra. Checkpointing strategies for shared high-performance computing platforms. *Int. Journal of Networking and Computing*, 9(1):28–52, 2019.

[30] A. Hori, K. Yoshinaga, T. Hérault, A. Bouteiller, G. Bosilca, and Y. Ishikawa. Overhead of using spare nodes. *Int. J. High Perform. Comput. Appl.*, 34(2), 2020.

[31] C.-h. Hsu and W.-c. Feng. A power-aware run-time system for high-performance computing. In *SC'05: Proceedings of the 2005 ACM/IEEE Conference on Supercomputing*, pages 1–1. IEEE, 2005.

[32] IBM Spectrum. LSF (Load Sharing Facility). `https://www.ibm.com/docs/en/spectrum-lsf/10.1.0`, December 2022.

[33] T. Jakobsche and F. M. Ciorba. Using Malware Detection Techniques for HPC Application Classification. In *Proceedings of The International Conference for High Performance Computing, Networking, Storage, and Analysis (SC24) Workshop on Cyber Security in High Performance Computing (S-HPC'24)*, November 2024.

[34] R. Jia, Y. Yang, J. Grundy, J. Keung, and L. Hao. A systematic review of scheduling approaches on multi-tenancy cloud platforms. *Information and Software Technology*, 132:106478, Apr. 2021.

[35] A. Kamatar, M. Gonthier, V. Hayot-Sasson, A. Bauer, M. Copik, T. Hoefler, R. Castro Fernandez, K. Chard, and I. Foster. Core Hours and Carbon Credits: Incentivizing Sustainability in HPC. *arXiv preprint arXiv:2501.09557*, 2025. `https://arxiv.org/abs/2501.09557`.

[36] B. Khemka, R. Friese, S. Pasricha, A. A. Maciejewski, H. J. Siegel, G. A. Koenig, S. Powers, M. Hilton, R. Rambharos, and S. Poole. Utility

maximizing dynamic resource management in an oversubscribed energy-constrained heterogeneous computing system. *Sustainable Computing: Informatics and Systems*, 5:14–30, 2015.

[37] L. Kury. *Automated Selection of Scheduling Algorithms for Parallel Scientific Applications Using Reinforcement Learning with OpenMP*. PhD thesis, University of Basel, Basel, Switzerland, 2022.

[38] LANL, NERSC, and SNL. APEX workflows. Technical Report SAND2016-2371 and LA-UR-15-29113, Sandia National Laboratories, Los Alamos National Laboratory and National Energy Research Scientific Computing Center, March, 17 2016.

[39] K. Leal. Energy efficient scheduling strategies in federated grids. *Sustainable Computing: Informatics and Systems*, 9:33–41, 2016.

[40] L. Lin and A. A. Chien. Adapting datacenter capacity for greener datacenters and grid. In *Proc. 14th ACM Int. Conf. on Future Energy Systems*, e-Energy '23, page 200–213. ACM, 2023.

[41] Y. Liu and H. Zhu. A survey of the research on power management techniques for high-performance systems. *Software: Practice and Experience*, 40(11):943–964, 2010.

[42] The Large Unified Modern Infrastructure (LUMI) Supercomputer, November 2024. `https://lumi-supercomputer.eu/lumi_supercomputer/`.

[43] M. Maiterth, G. Koenig, K. Pedretti, S. Jana, N. Bates, A. Borghesi, D. Montoya, A. Bartolini, and M. Puzovic. Energy and power aware job scheduling and resource management: Global survey—initial analysis. In *2018 IEEE International Parallel and Distributed Processing Symposium Workshops (IPDPSW)*, pages 685–693. IEEE, 2018.

[44] M. Maiterth, T. Wilde, D. Lowenthal, B. Rountree, M. Schulz, J. Eastep, and D. Kranzlmüller. Power aware high performance computing: Challenges and opportunities for application and system developers—survey & tutorial. In *2017 International Conference on High Performance Computing & Simulation (HPCS)*, pages 3–10. IEEE, 2017.

[45] S. Matsuoka, J. Domke, M. Wahib, A. Drozd, A. A. Chien, R. Bair, J. S. Vetter, and J. Shalf. Preparing for the Future—Rethinking Proxy Applications. *Computing in Science & Engineering*, 24(02):85–90, 2022.

[46] H. Menon, J. Diffenderfer, G. Georgakoudis, I. Laguna, M. O. Lam, D. Osei-Kuffuor, K. Parasyris, and J. Vanover. Approximate High-Performance Computing: A Fast and Energy-Efficient Computing Paradigm in the Post-Moore Era. *IT Professional*, 25(02):7–15, 2023.

[47] A. Mohammed, A. Cavelan, F. M. Ciorba, R. M. Cabezón, and I. Banicescu. Two-level Dynamic Load Balancing for High Performance Scientific Applications. In *SIAM Parallel Processing (SIAM PP 2020)*, Seattle, February 2020.

[48] A. Mohammed, J. H. Müller Korndörfer, A. Eleliemy, and F. M. Ciorba. Automated Scheduling Algorithm Selection and Chunk Parameter Calculation in OpenMP. *IEEE Transactions on Parallel and Distributed Systems*, 33:12, 2022.

[49] A. W. Mu'alem and D. G. Feitelson. Utilization, predictability, workloads, and user runtime estimates in scheduling the IBM SP2 with backfilling. *IEEE transactions on parallel and distributed systems*, 12(6):529–543, 2001.

[50] T. Özden, T. Beringer, A. Mazaheri, H. M. Fard, and F. Wolf. Elastisim: A batch-system simulator for malleable workloads. In *Proc. 51st Int Conf. Parallel Processing*, ICPP '22, 2022.

[51] J. Parente and K. Bethea. INCITE program awards supercomputing time to 56 projects to accelerate science and engineering research. *OLCF Website*, November 2022. `https://www.olcf.ornl.gov/2022/11/14/incite-program-awards-supercomputing-time-to-56-projects-to-accelerate-science-and-engineering-research/`.

[52] T. Patel, A. Wagenhäuser, C. Eibel, T. Hönig, T. Zeiser, and D. Tiwari. What does Power Consumption Behavior of HPC Jobs Reveal?: Demystifying, Quantifying, and Predicting Power Consumption Characteristics. In *2020 IEEE International Parallel and Distributed Processing Symposium (IPDPS)*, pages 799–809, 2020.

[53] T. Patki, D. K. Lowenthal, A. Sasidharan, M. Maiterth, B. L. Rountree, M. Schulz, and B. R. De Supinski. Practical resource management in power-constrained, high performance computing. In *Proceedings of the 24th international symposium on high-performance parallel and distributed computing*, pages 121–132, 2015.

[54] S. Prabhakaran. *Dynamic Resource Management and Job Scheduling for High Performance Computing.* PhD thesis, Darmstadt University of Technology, Germany, 2016. `http://tuprints.ulb.tu-darmstadt.de/5720/`.

[55] A. Radovanović, R. Koningstein, I. Schneider, B. Chen, A. Duarte, B. Roy, D. Xiao, M. Haridasan, P. Hung, N. Care, S. Talukdar, E. Mullen, K. Smith, M. Cottman, and W. Cirne. Carbon-aware computing for datacenters. *IEEE Trans. Power Systems*, 38(2):1270–1280, 2023.

[56] G. P. Rodrigo, P.-O. Östberg, E. Elmroth, K. Antypas, R. Gerber, and L. Ramakrishnan. Towards understanding HPC users and systems: a NERSC case study. *Journal of Parallel and Distributed Computing*, 111:206–221, 2018.

[57] A. Sergeev and M. Del Balso. Horovod: fast and easy distributed deep learning in TensorFlow. *arXiv preprint arXiv:1802.05799*, 2018.

[58] O. S. Simsek, J.-G. Piccinali, and F. M. Ciorba. Increasing Energy Efficiency of Astrophysics Simulations Through GPU Frequency Scaling. In *Workshops of The International Conference on High Performance Computing, Network, Storage, and Analysis (SC-W 2024)*, New York, NY, USA, 2024. Association for Computing Machinery.

[59] A. Tarraf, M. Schreiber, A. Cascajo, J.-B. Besnard, M.-A. Vef, D. Huber, S. Happ, A. Brinkmann, D. E. Singh, H.-C. Hoppe, A. Miranda, A. J. Peña, R. Machado, M. Garcia-Gasulla, M. Schulz, P. Carpenter, S. Pickartz, T. Rotaru, S. Iserte, V. Lopez, J. Ejarque, H. Sirwani, J. Carretero, and F. Wolf. Malleability in Modern HPC Systems: Current Experiences, Challenges, and Future Opportunities. *IEEE Transactions on Parallel and Distributed Systems*, 35(9):1551–1564, 2024.

[60] C. Trinitis and J. Weidendorfer, editors. *Co-Scheduling of HPC Applications.* Number volume 28 in Advances in Parallel Computing. IOS Press, Amsterdam, 2017.

[61] M. Turilli, M. Hategan-Marandiuc, M. Titov, K. Maheshwari, A. Alsaadi, A. Merzky, R. Arambula, M. Zakharchanka, M. Cowan, J. M. Wozniak, A. Wilke, O. O. Kilic, K. Chard, R. Ferreira da Silva, S. Jha, and D. Laney. ExaWorks software development kit: A robust and scalable collection of interoperable workflows technologies. *Frontiers in High Performance Computing*, 2, Oct. 2024.

[62] F. Yang and A. A. Chien. Large-scale and extreme-scale computing with stranded green power: Opportunities and costs. *IEEE Trans. Parallel Distributed Systems*, 29(5):1103–1116, 2018.

[63] A. B. Yoo, M. A. Jette, and M. Grondona. Slurm: Simple linux utility for resource management. In *Job Scheduling Strategies for Parallel Processing.* Springer, 2003.

[64] C. Zhang and A. A. Chien. Scheduling challenges for variable capacity resources. In *Job Scheduling Strategies for Parallel Processing*, pages 190–209. Springer, 2021.

[65] Z. Zhao and R. Hartman-Baker. Deploying checkpoint/restart for production workloads at NERSC. In *International Conference for High Performance Computing Networking Storage and Analysis*, 2020.

Part II

Infrastructures and Applications

Chapter 4

Managing Distributed Cloud Infrastructures for Sustainability

Wedan Emmanuel Gnibga, Anne Blavette, and Anne-Cécile Orgerie

Diverse computing paradigms are continuously emerging, transforming how a device interacts with data and completes tasks. These advancements offer functional benefits, like allowing even basic devices to run powerful applications efficiently by offloading demanding tasks to computing centers with available processing power and storage. This ease of task delegation has led to a boom in the number of users, more powerful and critical applications, and a constant drive to expand computing infrastructures to keep up with intensifying demands. However, this growing capacity and geographical distribution comes with a rising environmental cost. Some solutions are being developed to mitigate the environmental impact, including infrastructure optimization (hardware and design choices), finding dynamic ways for lower energy use (both in the hardware and software), using eco-friendly energy sources, etc. This chapter delves into the evolution of the computing landscape, the environmental challenges it presents, and the solutions being developed to address them.

The rest of this chapter is organized as follows: Section 4.1 presents the evolution of cloud (and edge) computing infrastructures, and their growing environmental footprint. Section 4.2 discusses how distributed infrastructures can be designed while limiting their environmental impact. Section 4.3 details approaches that aim to optimize Cloud infrastructure operations and their environmental footprint. Section 4.4 presents methods to perform holistic optimization in order to synchronize the Cloud workloads with power generation and to optimize heat rejection patterns. Finally, Section 4.5 concludes this overview of sustainable management strategies for distributed Cloud infrastructures.

4.1 Cloud Infrastructures: Evolution and Impacts

This section explores a taxonomy of computing concepts, their energy consumption and environmental impact.

4.1.1 From the Cloud to the Edge

The idea of harnessing remote service (i.e., computing power, storage, network) is not new. Cloud computing fulfills this vision on a grand scale, offering a powerful and adaptable way to access computing power through the Internet. Back in 1990, John Romkey's Internet-connected toaster showcased the early potential for devices to connect and exchange data over a network [3]. Since then, everything tends to be connected to the Internet, and user behavior has progressively changed to rely more on the Internet-based services. Connected objects were estimated at 20-50 billion in 2020 [3, 51]. Furthermore, new trends like generative AI and cryptocurrencies mining require massive computing resources and accelerate the Cloud expansion. According to the International Energy Agency (IEA), since 2010, the number of Internet users worldwide has more than doubled, while global Internet traffic has expanded 25-fold [54]. This surge in demand necessitates significant upgrades to infrastructures that include both the telecommunication networks and datacenters (DCs). As a response, there is a rise of large-scale datacenters, housing vast networks of compute nodes (CNs) and service nodes (SNs) – storage systems and infrastructure management nodes – and offering service to users across the globe. This infrastructure is known as fully centralized [2] and can be compared to a large library (datacenter) accessible from anywhere.

Many service providers are moving toward spreading their datacenters out across different locations, thus building distributed infrastructures [35, 42, 84]. This model contains a network of relatively smaller datacenters (nano- and micro-DCs) instead of relying on giant ones. Distributed datacenters offer several advantages. They bring processing and storage closer to users, dramatically reducing latency for applications like streaming and gaming. Spreading data across locations also creates a more resilient system, minimizing disruptions during outages. Additionally, new datacenters can be added in growing regions for targeted expansion.

Yet, the distributed datacenter model is not a replacement for centralized giants. These two models work together in a hierarchical system to provide the best of both worlds, but that constitutes a duplication of datacenters. In this partially distributed infrastructure [2], data processing happens closer to its source at edge datacenters for real-time needs, while the Cloud still plays a crucial role for tasks requiring vast resources or long-term storage. This is where comes the Edge-Cloud continuum [82]. One can imagine it as a library with a central archive (the Cloud) for rare books and extensive research materials. But, there are also smaller reading nooks (Edge datacenters) spread throughout the library for quick access to popular books or newspapers – closer to where people might need them.

4.1.2 Specifications of Edge Computing

The goal of Edge Computing is clear: bring services and processing power closer to the edge of the network. This offers attractive features like faster response times, improved reliability, and better security. In fact, the latency induced in a centralized Cloud service is usually higher than 100 ms and this is not applicable for use cases like real-time video analytics [6]. However, while the objective is clear, there is still a lack of consensus when it comes to defining the exact boundaries of this Edge and how it interacts with other concepts. For instance, on the one hand, it is often seen as interchangeable with Fog computing [70, 100], a concept that CISCO proposed in 2012 and defined as "*a paradigm that extends Cloud computing and services to the edge of the network*" [1]. On the other hand, others consider the Fog as a distinct approach (with regard to the hardware and technology employed) that is positioned as a layer between the Edge and the Cloud, handling more complex tasks than pure edge devices [22]. Other concepts have also emerged for distributing computing in space, taking it closer to the users. In 2014, the European Telecommunications Standards Institute (ETSI) [25] introduced Mobile-Edge Computing – that evolved later to MEC – which places the service directly within the radio access network (RAN).

An Edge infrastructure exhibits several key characteristics that differentiate it from traditional Cloud models [62]. Here is a breakdown of some of the most important ones. First, it offers a dense geographical distribution of the computing service with limited capacity. Moreover, as mobile devices increase, Edge computing becomes even more relevant. It also includes location awareness. This attribute allows mobile users to access services from the closest server to their physical location by employing various technologies such as GPS, or wireless access points. In addition, Edge computing works with a mix of different technologies. This means devices like phones, servers, and networks can come in all shapes and sizes – that is heterogeneity. This variety brings flexibility, but can also make it tricky for the whole system to function smoothly (interoperability issues).

In short, distributing computing resources enhances applications' performance. Unfortunately, this comes at a considerable environmental cost, on top of those already incurred in the information and communication (ICT) sector.

4.1.3 Energy consumption and environmental impacts

According to the European Commission [27], the ICT sector, encompassing broadly any device or application that allows people to access, store, transmit, and manipulate information, consumed 7% of global electricity in 2022, and that number is expected to surge to 13% by 2030. This translates into 3-5% of the world's carbon emissions [27], matching the impact of air traffic. Therefore, it is essential to analyze this consumption and carbon footprint and identify means to reduce them. We focus here on Cloud infrastructures.

Accounting for datacenters only, the energy consumption and carbon emissions remain substantial. According to the IEA [54], datacenters used an estimated 240-340 TWh of electricity globally in 2022, accounting for roughly 1-1.3% of the electricity used worldwide. While datacenters are becoming more efficient (i.e., require less energy to compute the same amount of workload as before), the amount of data they handle increases more rapidly – with an increase of 340% datacenter workload from 2015 to 2022 [54] – leading to an increase in energy consumption by 20-40% each year. The IEA's 2024 report on electricity [55] still forecasts that datacenters' energy consumption (including cryptocurrencies mining and AI) will keep growing in the near future, thus ranging between 620-1050 TWh in 2026. Per [55], "*This corresponds to an additional 160 TWh up to 590 TWh of electricity demand in 2026 compared to 2022, roughly equivalent to adding at least one Sweden or at most one Germany*".

Fossil energies (such as natural gas, oil) are currently the main sources used to supply these Cloud infrastructures at a global scale [54]. Thus, these infrastructures represent a source of GreenHouse Gas (GHG) emissions – around 1% of energy-related GHG emissions according to the IEA [54] – that "*will not reduce without major concerted political and industrial efforts*" [30]. However, the actual environmental impact of datacenters goes beyond GHG emissions.

These infrastructures also require water resources for cooling purposes and even for generating the electricity powering them. Yet, although water is often taken for granted, it is a finite resource under stress as the effects of global warming intensify. In fact, "*only 2.5% is fresh and only 1% of that is accessible for direct human use; furthermore the water system is closed, i.e. there is a finite quantity in the ecosphere. Population growth and related land occupation and use, agriculture and irrigation and industrialization mean that demand is continually increasing*" [102]. David Fitzgerald estimated that "*a medium-sized datacentre (15 megawatts) uses as much water as three average-sized hospitals, or more than two 18-hole golf courses*" [29], which, extrapolated on a global view, is substantial and accelerates the water's unavailability. Moreover, David Mytton [76] estimated the global water use of datacenters. In a country like the US, the datacenter industry directly or indirectly draws water from 90% of US watersheds – among which 20% are part of highly stressed watersheds in the western USA – ranking it in the top 10 of water-consuming industry [88]. However, it is important to mention that the US Environmental Protection Agency predicted in 2023 that at least 40 US states will experience water shortages by 2024 [99]. This example reflects the global water stress.

To reduce this environmental impact while seeking service performance, one needs to identify some flexibilities in the system, i.e., some functionalities and levers that can be exploited at different layers of the datacenters to mitigate this impact. In the following, we dive into these flexibilities and explore recent studies that exploit some of them to improve infrastructures' performance and environmental footprint.

4.2 Building Efficient and Sustainable Cloud Systems

This section describes how a distributed infrastructure can be designed to achieve expected levels of performance while mitigating to some extent its nonetheless significant environmental impact. The infrastructure design can be optimized for both economic and environmental benefits. We focus here on the most widely studied environmental impacts of Cloud infrastructures in literature, namely electricity consumption, carbon emissions and water consumption.

4.2.1 Datacenter location and service deployment

Many candidate locations may host datacenters (large-scale or at the edge). However, they may not lead to an optimal or a sustainable operation. Thus, several authors propose intelligent placement schemes. Picking a good location for a large-scale datacenter is not just about space. Many factors come into play, such as land availability and cost, construction expenses, local electricity prices and mix, local average temperatures, and even water needs and availability for cooling [41, 88].

Similarly, deciding where to locate servers (or mini datacenters) in the Edge layer also involves balancing several competing goals like access delays, energy consumption and profit [9]. The server placement can even be made dynamical to follow the evolving user demands and telecommunication network patterns [104]. Edge datacenters handle diverse applications that may not be deployed on all the nodes because of the capacity limitations. Hence, one may additionally need to strategically distribute the applications across the servers in each location [12].

Selecting a strategic location is an important step toward achieving high performance in terms of service, energy consumption and environmental impact. In addition, the choice of cooling design also determines the performance of the datacenters, especially their energy and water consumption.

4.2.2 Cooling system design

Thermal management is essential for a datacenter to maintain the optimal temperature, humidity, and airflow levels for its equipment to operate reliably and efficiently. There are many cooling system designs available, differing by technology, energy and water consumption profiles: air free-cooling, mechanical cooling, and immersion cooling.

Air free-cooling does not require water to operate. It allows to use naturally cool air to lower the temperature in the datacenter, without the need for mechanical refrigeration. This technique is broadly used in research projects [11, 89]. However, it is only applicable in specific meteorological conditions: locations with colder temperature (<20°C) and higher humidity (>60%) [89]. Therefore, many locations are not suitable for this design. Even regions with average satisfactory weather conditions may have unsuitable periods during the year. Unfortunately, free-cooling will become less applicable as the global climate warms. Moreover, an exclusive use of free-cooling in large datacenters could be inefficient and challenging to deploy, as air has a low thermal conductivity.

A second kind of cooling design is the mechanical cooling system. Server rooms can be cooled down using the classical air conditioners. However, they are energy-demanding, usually causing a high carbon impact. Alternatively, in large-scale datacenters [13, 108] (at the scale of megawatts), the generated heat is conveyed by water through mechanical refrigeration processes and dissipated into the atmosphere via giant cooling towers. Two heat dissipation technologies can be employed in the cooling towers [39]. On the one hand, they may rely on a water state change principle (i.e., convert water from liquid to vapor state and release it to the atmosphere), a technique called evaporative cooling. On the other hand, they may be based on closed heat exchangers operating without water loss. This cooling approach is known as dry cooling. While dry cooling saves water, it also consumes a significant amount of energy, as opposed to evaporative cooling. Nevertheless, regardless of the type of mechanical refrigeration used, it is often associated with free-cooling, like in Amazon Web Service (AWS) datacenters [5], to take advantage of cheap and eco-friendly cooling when the weather is favorable.

Major service providers are investing in the next generation of cooling technologies to reduce or compensate their water and energy consumption. For instance, Microsoft [90] and Google [16] plan to be water-positive by 2030, i.e., to release more water than their datacenters' average consumption per year. Their proposed strategy would rely on two ways: reducing water use intensity (water consumed per megawatts) and replenishing water in water-stressed regions. As for the energy consumption, it is tackled by the progressive experimentation of less energy-demanding cooling technologies like immersion cooling [83]. Immersion cooling consists in bathing datacenter equipment in an electrically non-conductive dielectric fluid, whose thermal conductivity is far superior to that of air. Hence, the heat is transferred to that liquid in order to be dissipated in the surrounding environment. In 2018, Microsoft even initiated a research project to determine the feasibility of subsea datacenters [96], which would eliminate any mechanical cooling system.

Addressing the reduction of energy and water consumption when designing and building a Cloud infrastructure is a step to reduce its environmental impact. However, besides reducing the environmental impact during the design phase, another complementary approach consists in reducing this impact during the operational phase. In the following, we describe how to optimize datacenters' operation in order to reduce their footprint.

4.3 Optimizing Datacenter Management and Footprint for Sustainability

This section details dynamic approaches for optimizing computing resource utilization and minimizing the environmental impact of an operating infrastructure only – the optimization opportunities offered in the electrical system (including using renewable energy sources) are discussed in Section 4.4. We first explore metrics and measurement tools to assess environmental impacts in Cloud environments. Then, we examine literature that aims to optimize Cloud management based on these metrics, on one hand using hardware-level optimizations, and on the other hand exploiting software-level optimizations.

4.3.1 Metrics and measurement tools: a prerequisite

Optimizing a datacenter infrastructure, especially in a context of Edge computing, requires to define some critical metrics to be considered. We structure the analysis into three main topics: performance metrics, electrical metrics, and environmental metrics.

In terms of performance, users are primarily concerned with the Quality of Service (QoS), which focuses on the technical performance of the service and ensures that it operates according to defined criteria. In the literature [10, 62], the QoS is generally assessed using indicators such as: (1) Service availability, which is the percentage of time a service is operational and accessible to users. A high availability guarantees minimum downtime and constant service. (2) Latency accounts for the amount of time it takes for data to be transmitted from one point to another, typically from the user device to the remote server and inversely. Low latency is essential for real-time applications such as video analytics and online gaming. In the context of Edge computing, latency is often considered as the main metric to optimize the service delivery. (3) Throughput is the amount of data a network can transfer over a given period of time. High throughput is essential for applications involving large data transfers, such as file downloads and video streaming. A contract called service-level agreement (SLA) is often involved to define the expectations of both the service provider and the users. It describes the acceptable performance levels and the consequences of non-compliance. Green-SLA goes a step further by including clauses that encourage eco-friendly practices from the providers' side. Alongside the evaluation of technical characteristics, one may also estimate the Quality of Experience (QoE) [10], which reflects the user's perception of a service's performance. It is a subjective notion that takes into account factors like ease of use, reliability, security, etc. The QoS contributes to the QoE in some critical applications.

From an electrical point of view, the literature also provides key indicators to assess an infrastructure's performance. (1) Power demand (in watts) is the instantaneous amount of consumed electricity. This metric, along with the energy consumption described below, is particularly crucial to size optimally the infrastructure's electrical system and to sign a contract with an electricity supplier. (2) Energy consumption (in watt-hours) measures the total amount of electricity consumed by the infrastructure throughout a given time horizon, typically a day, a month, a year, etc. It is generally expressed in gigawatt-hours for large-scale datacenters. (3) Power Usage Effectiveness (PUE) is the ratio between energy used by the IT equipment and the overall energy consumed in a datacenter.

Two major physical metrics are often regarded in environmental impact accounting: greenhouse gas emissions and water consumption.

Greenhouse gas emissions: They capture the greenhouse gases emitted by an infrastructure over its entire lifespan. However, greenhouse gases are multiple – carbon dioxide (CO_2), methane (CH_4), nitrous oxide (N_2O), fluorinated gases, etc. – with different global warming potential levels. Hence, the total greenhouse gas emissions are often simplified as an equivalent carbon footprint, expressed in grams of CO_2 equivalent (gCO_2eq). As an illustration, in 2022, CO_2 accounted for 80% of all US greenhouse gas emissions from human activities [98]. This metric can also be normalized as the carbon intensity of the infrastructure, expressing the quantity of greenhouse gases emitted per unit of service (gCO_2eq/unit of service). The Greenhouse Gas (GHG) Protocol [34] is widely used as a standard to help organizations measure their carbon footprint. For instance, the authors of [18] state that big tech companies like AMD, Apple, Google, Huawei, Intel and Microsoft all rely on this protocol in their annual sustainability reports to track their impact. The GHG Protocol categorizes emissions into three scopes: Scope 1 (direct emissions), Scope 2 (indirect emissions from purchased energy), and Scope 3 (upstream and downstream supply chain emissions).

Scope 1 emissions come from direct sources controlled by the firm itself. These include on-site fuel combustion (diesel, gasoline, etc.) for machinery (including backup electricity generation) and company-owned transportation, refrigerants used in buildings, and the use of chemicals. Scope 2 emissions originate from the energy purchased, including grid electricity and energy for heating the offices. These emissions are highly dependent on the amount of energy consumed and the type of energy sources used (e.g., renewable or fossil fuels). Many works [18, 56, 80] usually base their carbon accounting on scopes 1 and 2. See also Chapter 5 for more details about the GHG protocol.

To account for scopes 1 and 2, the energy consumption of datacenters is measured through power-meters typically embedded in Power Distribution Units (PDUs). These measurements can be refined by hardware sensors directly plugged into servers (e.g., on the Baseboard Management Controller) or into processors (like, for instance, Running Average Power Limit tools). Although physical energy meters are accurate, they require additional hardware and software configuration. Furthermore, they do not provide a breakdown of energy consumption for a specific running application instance, nor for specific data being transferred across the network. Hence, several models and tools called "software power meters" are emerging to estimate the energy consumption at different levels of granularity. For server energy estimation, one may resort to tools like PowerAPI [28], Likwid [97] and Scaphandre [52]. They are based on the Intel CPU RAPL [23], a hardware technology that uses high-frequency registers (1000 Hz) to estimate a processor's power consumption. The authors of [57] conduct a comprehensive experimental comparison of these software-based power meters and others.

Finally, scope 3 emissions come from all other activities, including the full upstream and downstream supply chain. They often comprise the embodied carbon associated with the materials, manufacturing, and transportation of equipment (servers, cooling systems, network devices). In practice, this scope is complex to assess and is overlooked by Cloud providers.

Water consumption: This metric accounts for the amount of water used by the infrastructure (including electricity generation, cooling and humidification) over a time horizon. It is often expressed in units of liters (L), cubic meters (m^3), or gallons (gal). It is worth noting the distinction between the water withdrawn and that consumed (dissipated into nature) by the datacenter. We focus on dissipated water. In practice, one uses the Water Usage Effectiveness (WUE), that depicts the amount of water used per unit of energy consumed by the IT equipment (L/kWh), to evaluate the water usage of an infrastructure. We propose in [39] a detailed analysis of the water consumed in a datacenter, both due to cooling and to powering the IT equipment (e.g., for electricity generation), depending on the Cloud workload, the local weather conditions, the local electricity mix and the cooling system. In this study, we examine the two most widespread cooling systems, namely dry cooling and evaporative cooling (aka, adiabatic cooling) and we express trade-offs between water consumption, power consumption, carbon footprint and financial costs. Chen *et al.* [19] also propose a quantification of the on-site (i.e., for cooling purposes) and indirect (i.e., water used for generating the electricity consumed by the datacenter) datacenter water consumption. The authors proposed the Water Scarcity Usage Effectiveness (WSUE) as a sustainability metric that captures the impacts of water consumption on regional water availability. A more extensive selection of green metrics for analyzing individual key performance indicators of datacenters is detailed in [81].

GHG emissions and water consumption are only two environmental indicators impacted by Cloud infrastructures among numerous ones (e.g., ozone depletion, human toxicity, acidification, land use, eutrophication). However, these other indicators are not yet estimated by Cloud providers, and literature is scarce on this subject.

4.3.2 Hardware-level energy optimizations

Besides the efforts undertaken by manufacturers to improve the energy efficiency of the computing equipments themselves, they also provide hardware features that can adapt their properties based on performance needs, further optimizing energy consumption, and consequently GHG emissions.

Modern servers are increasingly designed to dynamically scale their performance in real-time based on the needs of the applications they are running, in alignment with the QoS requirements. Most modern processors (Intel x86, AMD, and ARM) support the DVFS technology (Dynamic Voltage and Frequency Scaling, also known as CPU throttling), which is widely exploited for saving energy through decreasing the supply voltage and frequency of computing units (e.g., processors), and has been proven to be efficient [109]. For instance, [61, 67] propose algorithms that considers how long tasks can wait (slack time) before their completion deadline, to scale down the performance of Edge servers, thus reducing their energy consumption (by 13.76% in [61]). To implement DVFS, the processor needs to have a voltage regulator (VR) that can control the supply voltage, and a frequency generator that can control the clock frequency. In this setup, DVFS decides how much power the processor needs, and the VR adapts the voltage, and therefore the power, precisely and efficiently. Using these hardware technologies, the authors of [109] propose an approach based on convex optimization to find the best settings for both the power management of each CPU core (DVFS) and the associated voltage regulators. They achieve a total energy reduction of the system by 22.4% with negligible QoS overhead. However, DVFS increases the computing time of running tasks.

When the processors have no workload to process, one may exploit C-States (CPU Sleep States) to put them into various low-power states. In this architecture, C0 corresponds to the active state, C1 to the idle state and C-states beyond C1 are deeper sleep states. Based on this feature, the authors of [45] experiment with the potential power savings of different sleep states on Intel x86 servers and find around 80% processor power saving when the servers are idle. The authors of [7] also consider this opportunity to reduce energy consumption. However, the technique degrades the performance of their considered applications (microservices with 30-250 μs latency tolerance) due to the long transition latency (of the order of 100 μs) to wake up from deep sleep states. Therefore, they propose AgilePkgC, an architecture based on the C-states model, with improved start-up time and extension to other I/O hardware (e.g., memory, PCIe bus). They achieve an average 25% energy reduction on Intel Skylake-based servers [7].

One may also use features like Power Delivery Throttling (or power capping) that rather aims to limit the maximum power delivered to the processor.

In this configuration, the cap has no effect until the server reaches its power consumption limit. At that point, a management processor adjusts CPU P-states (performance states) and clock throttling to limit the power consumed. On Intel servers, this feature can be set using the RAPL technology. The authors of [106] use power capping to lower the energy consumption of servers. Many other server features exist to reduce server consumption and are listed in [58, 74].

Edge infrastructures handle a wide range of services, each with specific computing and energy profiles. While hardware features mentioned previously offer benefits, further energy savings can be achieved by using specialization in the equipment design. Some servers have two or more of the following: CPUs (for standard applications), GPUs (artificial intelligence-based applications), and specialized chips such as field-programmable gate arrays (FPGAs). The idea consists in providing hardware of various capacities and using the most suitable one for a given application. For instance, in 2012, Intel released a CPU-GPU-FPGA architecture [53]. Besides incorporating heterogeneous chips in infrastructure equipment, heterogeneous servers are often used within and across edge datacenters. Low-power devices like Raspberry PIs can participate in an infrastructure design with regular servers, laptops, and workstations. This creates a system where applications can be scheduled to run on the most appropriate device based on their resource needs and the nodes' energy efficiency.

In summary, servers can be configured and dynamically adjusted to some extent to use less energy and reduce their carbon footprint. Additionally, a variety of software techniques have emerged to meet service level agreements, while reducing the energy consumption and resulting carbon footprint.

4.3.3 Resource management optimizations

It is worth understanding how applications are running in a datacenter. Just like online service providers do, a wide range of research work is based on virtualization [4], containerization technologies like Docker, and unikernels [73] to isolate user applications from each other and the underlying hardware resources they run on. These technologies guarantee resource availability to each user application, and therefore compliance with targeted QoS. Interestingly, they also provide great flexibility since they all support migration from one node to another. This is especially useful in the context of mobility and for dynamically adjusting the resource usage to achieve a better QoS [95]. Besides virtualization, critical but resource-intensive applications, suited to be deployed on the Edge layer, can be decomposed into smaller, independent units called microservices [79]. This modular approach allows for flexible deployment across edge resources, ensuring high scalability of the applications. In the rest of this chapter, we identify a task or a job as a computing workload (either a full application or a microservice instance), running in a virtual environment, i.e., a virtual machine (VM) or a container.

Task scheduling (i.e., when to execute a given task) and resource allocation (i.e., where to execute it) are joint activities composing a resource management policy. Task scheduling approaches are influenced by the characteristics of the tasks to be run. In that sense, batch jobs, defined by deadlines, allow us to defer their execution and potentially interrupt them as long as the deadline is met. Conversely, service jobs require immediate resource allocation upon arrival. Furthermore, the scheduling process needs to account for potential dependencies among the tasks. In fact, certain tasks may rely on the completion of others, so the schedulers must incorporate these relationships in decision-making. Task scheduling can be made offline or online, according to the quantity of information available on the tasks. In offline scheduling, the scheduler has a complete knowledge of task characteristics (e.g., resource requirements, execution duration, deadline, etc.) and makes decisions prior to their execution. With online scheduling, the scheduler operates with limited task information. Hence, decisions occur at runtime, based on the characteristics of the currently executing tasks and the system state. Finally, schedulers can be centralized or distributed. In the context of Edge computing, centralized schedulers have an overall knowledge of the infrastructure to balance the load on the suitable resources, yet providing limited scalability in large infrastructures. As for distributed resource scheduling, each site is provided with a scheduler capable of coordinating with the others through peer-to-peer communications, with the inconvenience of losing the global knowledge of the system. These approaches are surveyed in [70].

Most resource management algorithms are rule-based (heuristics) and aim to accurately distribute the jobs on the available computing resources in order to optimize their utilization. Resource allocation is a bin packing problem. The literature provides popular algorithms to solve it. For example, the Round Robin (RR) policy places each task on the next available host based on the concept of circular queue. The Max Fit algorithm places them in the servers that have the maximum available resource. Conversely, the Best Fit (BF) policy places the tasks into the servers that have the minimum available resources and that can accommodate them. The following surveys provide extensive lists of resource allocation algorithms [50, 105].

Meta-heuristics are another approach to approximate solutions for resource management. These fall into two main categories: (1) evolutionary algorithms like genetic algorithms [24, 103] (inspired by the process of natural selection), and (2) swarm intelligence algorithms like particle swarm optimization [85, 87], which is a method inspired by the behavior of bird flocks.

Beyond heuristics, resource management often relies on other powerful approaches: game theory and machine learning. Game theory is a theoretical method used to model interactions between multiple tasks competing for the limited resources. By analyzing these interactions and potential outcomes, game theory can guide resource allocation strategies that promote service efficiency, as explored in [49, 93]. Machine learning algorithms analyze historical data on resource usage, application demands and network conditions. Then, based on that, they forecast future resource requirements and make dynamic allocation decisions that optimize resource utilization and application performance [86, 94].

The techniques described above are conservative, as they estimate the free resource capacity on a server by taking the difference between its physical capacity and the sum of peak resources reserved by the tasks running on it. This approach protects against the worst case scenario where all tasks use their maximum allocated resources at the same time. In practice in Cloud environments, tasks rarely (if ever they do) use their maximum allocated resources. Therefore, the sum of tasks' resource usage at a given time is consistently less than the sum of their allocated resources. This gap creates an opportunity for application collocation, making it possible to run more tasks on a server without overloading it (i.e., not degrading the QoS requirements). This resource management is known as overcommitment. Several resources can be overcommited, such as memory and CPU. Yet, the performance of overcommitment techniques highly depends on the workload (type of tasks, variability of resource usage patterns, etc.). The authors of [14] show that overcommitment policies can improve machines' usable CPU capacity by 10-16% compared to no overcommitment. Moreover, in [43], the authors show that physical memory overcommitment may be effective for doubling the number of launched tasks. However, memory overcommitment is a complex task, as it is difficult to predict how many additional tasks can be launched before performance is adversely affected. Moreover, memory overload can easily lead to the whole server crash.

While management strategies often aim to trade off various objectives, some mechanisms specifically target energy efficiency.

4.3.4 Energy-aware mechanisms

A key strategy for minimizing energy consumption is dynamic resource consolidation. It utilizes live migration to transfer tasks among physical servers in order to improve the utilization of resources. Dynamic consolidation is typically useful when some tasks work together to deliver a service. Therefore, periodically revisiting the resource allocation to collocate such tasks on the same server allows to share the data source (e.g., memory or disk) or to reduce data communication delays and the associated energy consumption. Interestingly, a new use case of resource consolidation arises, associating it with on/off policies to further improve datacenters' energy efficiency. This consists in grouping workload on a reduced number of active servers, and powering down, or putting in deep sleep mode, the idle ones. This approach reduces the overall static energy consumption and maximizes the utilization of active machines.

Many optimization strategies are proposed in the literature to solve the problem of resource consolidation. We do not provide here an exhaustive list of these strategies, but we give an overview on some of the most popular approaches identified in surveys. For instance, in the Modified Best Fit Decreasing (MBFD) algorithm, the tasks (VMs) are sorted in decreasing order of their current CPU utilization, and each task is allocated to the server that provides the least increase of power consumption due to that task [15]. Moreover, the allocation is periodically revised in a context of consolidation, by choosing the most power-efficient nodes to execute tasks. Other algorithms like worst-fit decreasing (WFD) and Almost Worst-Fit Decreasing (AWFD) algorithms are developed. An extensive list of such algorithms is provided in [50, 75, 107]. Yet, most of these approaches do not consider network congestion issues and may thus lead to poor energy-efficiency in practice as we have shown in [36].

Recent research works are also exploiting the idea of consolidation. For instance, the authors of [72] design the Workload Aware Virtual Machine Consolidation Method (WAVMCM), a virtual machine consolidation scheme in small-scale storage-as-a-service datacenters. Their experimental results show that the proposed method reduces the number of active servers by 9% compared to probabilistic algorithms based on artificial intelligence, and it saves 15% in energy consumption. In the same direction, the authors of [77] propose a predictive, online and distributed algorithm for the scheduling of computing jobs, which dynamically reorder them to achieve energy efficiency.

Overcommitment can be also exploited to reduce the number of active nodes, but that may require a trade-off with the QoS [69], for critical applications. Moreover, one may delay batch tasks (temporal load shifting) to periods when their execution demands the least power. This strategy allows to free servers when it is possible and to switch them off. Finally, in distributed datacenters, an extensive consolidation policy can be adopted by migrating the load across them (spatial load shifting) and achieving an overall optimization of energy consumption [36].

Another way of saving energy in a datacenter is through performance adjustment (i.e., self-configuration and adaptation) of the running applications. For instance, the authors of [59] propose a system that enables different frequency scaling policies for different application classes (i.e., latency-sensitive or best effort applications). The system is deployed in production at Google's datacenters and improves power savings by 9% on average. The authors of [71] have also developed Energy-Efficient Co-Locating and Self-Tuning (ECoST), a method for optimizing energy efficiency and self-tuning for data-intensive workloads. Considering the video streaming application, the authors of [8] develop an analytical framework that can provide the users with an estimate of how much energy they can save by choosing to view a lower quality stream of the video. The framework takes as input the network conditions and characterization of the video to be viewed (slow versus fast motion, resolution), and yields as output the energy savings with different resolutions of the video to be viewed. Then, the users can make a decision on the version of the video to view. However, this approach focuses on the user-side, which is not dynamically tuned.

4.3.5 Validation methods

All the contributions described above adopt various methods to validate their work. Some authors validate their strategies using benchmarks [14, 43]. This consists in deploying the algorithms on a real experimental or production environment and measuring their performance. Although this option highlights the exact behavior of the solutions, it may not be accessible for large-scale deployments (i.e., they are costly to build) or not recommended for high-risk operations (e.g., system crash). Therefore, it is common to simulate the operation conditions using software tools that implement mathematical models of objects like servers, network devices, etc. The models may require a calibration using parameters that describe specific conditions (e.g., energy consumption of specific servers). Hence, these parameters are provided using real measurements (a practice to encourage) or are based on vendor indications. The studies based on simulation fall into two categories with regard to the workloads used: (1) real-world traces (or data sets) and (2) generated workload. Real-world traces [50, 94, 105] are collected in real datacenters (e.g., Google clusters) and help to make offline analysis based on realistic conditions. However, these traces are generally collected for a short duration (e.g., one month) and do not enable long-time (e.g., annual) and/or large-scale analysis. Secondly, they depend on specific applications and users, and may not reflect task patterns in certain use cases. Therefore, some authors resort to workload generation [61, 67, 103], based on a statistical analysis that describes the task patterns (e.g., arrivals, duration, resource usage, etc.).

Table 4.1 summarizes the optimization methods described above and their characteristics.

Resource management can help to reduce the datacenters' footprint. Yet, other means allow to improve environmental indicators. Indeed, the electricity mix powering datacenters varies over time, and dynamic optimization strategies exist to take advantage of this variability in order to improve both carbon and water footprint of datacenters. Moreover, cooling systems represent a highly consuming element of datacenters both in terms of electricity and water. In the following, we explore strategies targeting these two parts (e.g., the electrical system and the cooling system of datacenters).

4.4 Datacenter Flexibility and Synergy with Power Generation and Grid

Datacenter operation can be co-optimized with power generation and grid dynamics, reducing negative environmental impacts. This approach draws on datacenter load flexibility, which can arise from self-consumption, cooling

Table 4.1: Strategies employed to optimize infrastructure operation.

References	Considered criteria					
	Energy saving	Resource management	Consolidation + on/off	Overcom-mitment	Perf. adaptation	Workload type
[61, 67]	✓	Heuristics	×	×	DVFS	Generated
[50, 105]	✓	Heuristics & AI & meta-heuristics	✓	×	DVFS	Generated & real traces
[24, 85, 103]	×	Meta-heuristics	×	×	×	Generated
[87]	✓	Meta-heuristics	✓	×	×	Real traces
[93]	✓	Game theory & heuristics	×	×	×	Generated
[49]	✓	Game theory & heuristics	×	×	×	Real traces
[86]	✓	AI	×	×	×	Generated
[94]	✓	AI	×	×	×	Google traces
[14]	×	Heuristics	×	✓	×	Google traces & benchmarks
[43]	×	Heuristics	×	✓	×	Benchmarks
[72, 75, 77, 107]	✓	Heuristics & AI & meta-heuristics	×	✓	×	Generated & real traces
[69]	✓	Heuristics	✓	✓	×	Generated & real traces
[71]	✓	AI	×	×	App. perf. tuning	Measurement & simulation
[8]	✓	Heuristics	×	×	Slow-motion streaming	Measurement & simulation
[36]	✓	Heuristics	✓	×	App. perf. tuning	Measurement & simulation

flexibility, or compute load flexibility. We first present works targeting the energy management of power sources and loads, as well as energy storage units, to manage computing workloads in a sustainable and efficient way. Then, we explore the notion of self-consumption, which incentivizes the synchronization of power consumption with local power generation usually based on renewables. Finally, we detail approaches that create and exploit datacenter flexibility arising from cooling infrastructures.

4.4.1 Synchronizing workload with renewable energy

User tasks fall into two categories: inflexible tasks that must be performed as soon as submitted, and batch jobs that can be delayed provided that the expected quality of service is fulfilled. Therefore, in infrastructures powered with renewable energy, batch jobs may be deployed according to the abundance of electricity production. Based on this observation, GreenSlot [110] is a scheduler that predicts the amount of solar energy of the near future and schedules the workload to maximize the renewable energy consumption while meeting the jobs' deadlines. When the solar generation is not sufficient, it selects the time when the external grid energy is the cheapest to submit the tasks with respect to the deadlines. The authors of [44] propose an approach for scheduling batch jobs with due date constraints taking into account the renewable energy's availability to reduce the use of grid-imported energy. Some more recent contributions use machine learning techniques to automatically learn effective job scheduling policies by continually adapting them to the datacenters' complex dynamic environment (computing and renewable energy resources) [101], while others extend the scheduling policy exploration to service tasks so as to consume more on-site renewable energy [48].

In distributed infrastructures, each datacenter delivers a local service and may participate in a large ecosystem of datacenters via electrical or telecommunication interconnections. In non-interconnected or weakly connected electrical networks suffering from congestion issues, such service providers can exploit the temporal variations in on-site productions by routing the load to a datacenter with available computational resources and renewable electricity. This is called energy-aware load migration [38]. However, within a well-interconnected electrical network, energy-aware load migration (i.e., transferring one workload from a renewable energy-poor datacenter to a renewable energy-rich one) may not increase the overall renewable energy consumption of a power system. The excess of renewable production at one datacenter may indeed be absorbed by other surrounding consumers, given the usually low-power consumption of edge nodes, and their correspondingly low level of on-site power generation, compared to the total consumption from the surrounding neighbors. Inversely, its deficit at another datacenter is met by closer power plants, including renewable power plants. In case a task is migrated to a renewable-rich datacenter, the consumers powered by the excess of renewable prior to the migration would also need to be powered from additional power generation means. However, consuming power from local sources may qualify this as self-consumption, which benefits from economic incentives in numerous countries in order to incite consumers to better synchronize their consumption with their local production. Self-consumption may also reduce power line losses (which may or may not be compensated by the additional energy required to migrate tasks though) and electrical congestion issues. The concept will be described in Section 4.4.2. For instance, the authors of [63] propose an algorithm that predicts solar and wind power availability in interconnected datacenters and migrates virtual machines to the datacenters with the most renewable energy. The authors of [17] propose algorithms that take advantage of this cooperation between datacenters to synchronize their on-site photovoltaic production and their consumption by migrating virtual machines to the most abundant sites or using a proposed tarification scheme mutualizing the production and consumption over all the datacenters.

The same idea of planning tasks to be executed when (temporal load shifting) and where (spatial load shifting) energy is less carbon-intensive can be applied in infrastructures without renewable energy sources (on-premises). For the same reasons as mentioned earlier, spatial load shifting is particularly relevant for non-interconnected grids. Also, as the energy mix of grids fluctuates over time, some research works use their carbon intensity as an input for decision-making, reducing the carbon impact by performing temporal load shifting [92].

As service tasks cannot be fully synchronized with renewable energy production and since batch jobs can only be shifted to a certain extent, the renewables are often associated with Energy Storage Systems (ESSs) to store energy from the timescale of seconds (short-term storage) to months (seasonal storage). Hence, the datacenters consume more renewable energy, and become more independent from the grid [37, 66]. For instance, in the green datacenter prototype Parasol [40], whose power supply is composed of a set of solar panels, a battery and a grid-tie, the authors experiment a scheduling approach called GreenSwitch that dynamically manages the workload and the energy sources. This combination of supply-side and demand-side mechanisms optimizes the overall cost of electricity, the battery usage, and the quality of service. Finally, the authors of [65] propose a two-layer hierarchical energy management algorithm for a datacenter powered by an association of wind, tidal, battery and hydrogen energy storage systems. The strategy uses a genetic algorithm for day-ahead scheduling (first layer) and real-time dispatching with rolling predictive control (second layer).

The above-mentioned approaches involve an energy mix including potentially non-renewable energy and require the datacenter to consume electricity from the electrical grid. Now, some new research directions arise and focus on autonomous datacenters. For instance, the authors of [78] study a datacenter powered by 100% renewable energy without any contribution from the grid. The renewable power supply is composed of photovoltaic panels and wind turbines and is associated with batteries and a hydrogen storage system. Both the electrical and the computation parts of the datacenter have independent optimization mechanisms, and a negotiation protocol is introduced to match the power consumption with the power generation. In the same direction, the authors of [64] propose an energy management and sizing strategy for a 1 MW standalone green datacenter powered by a hybrid tidal and photovoltaic system associated with a battery and a hydrogen storage system. These prototype datacenters can be easily expanded in a distributed environment, but they require some additional space for energy storage systems compared to traditional datacenters.

Datacenter operators are investing in renewable energies, while trying to synchronize their load to the renewable energy availability to some extent, with the help of energy storage systems, thereby avoiding reliance on polluting sources. Yet, relying only on renewable energy sources may be prohibitive in terms of investment costs for energy production means (e.g., wind turbines, photovoltaic panels) and energy storage systems (e.g., batteries, hydrogen storage). The trade-off between grid dependency and autonomy costs is extensively analyzed in [37].

In this section, the detailed approaches all aim to produce and consume on-site energy. However, datacenters can also be involved in collective self-consumption programs, enabling them to synchronize their consumption or production with surrounding consumers or producers, under certain conditions. The concept of self-consumption is detailed in the next section.

4.4.2 Notion of self-consumption (on-site generation)

Worldwide, individuals and businesses (like ICT companies) can set up their own renewable energy systems (generally solar panels) to power their facilities (self-consume). However, their consumption may not be fully synchronized with on-site power generation. One alternative may consist in adding an energy storage system that can absorb the excess of production to release when the consumer needs it. However, the economical benefits of consuming a "free" electricity (once the infrastructure investment is done) may not compensate the cost of energy storage [37]. In case the excess of power is injected into the grid, several options exist.

First, the electricity can be purchased by an energy provider usually at a lower price than the price at which it is sold to the consumer. Another alternative consists in creating an energy community [26]. Energy communities are supported by specific legislation and rules designed to make energy exchanges economically advantageous for all members. In essence, participants are allowed to sell the electricity produced on their premises to other participants, at a tariff that is advantageous to all of them. Hence, this incentivizes a better synchronization between power generation and consumption at the level of the energy community, while benefiting from an aggregation effect. Under certain conditions, which may vary from one country to another, the energy consumption of energy communities may (partly) be qualified as collective self-consumption (CSC). Under this scheme, the electricity assumed to be self-consumed is paid at a lower tariff (e.g., through a reduction of network fees).

For instance, Denmark, Finland and Sweden allow CSC, but only on a limited scope (e.g., via private networks) or tolerate it within a regulatory framework presented in [32]. France also launched its CSC initiative in 2016, under government regulation [31]. A broader list of European countries that adopt the concept of CSC is surveyed in [33]. In the existing rules, energy is considered as self-consumed when it is produced and consumed within the boundaries of the energy community at the same instant, enabling members to benefit from the economic payoffs.

When CSC involves diverse participants (like households), the energy management strategies to adopt must consider the different possible billing methods. In this sense, several schemes are developed to allocate the aggregated power generation to each participant and share the incomes [21, 91].

As per most of the national regulations, the eligibility to CSC schemes is restricted to participants with a limited geographical dispersion. This may be justified by the fact that local CSC may reduce power losses and congestion issues significantly, which may not be the case if this concept is applied at a larger scale. However, its application at a larger scale may still lead to the benefit of better synchronization between energy generation and consumption, as we have shown in [38]. We therefore study a scheme enlarging the CSC applicable perimeter and evaluate its gains in the context of a distributed edge-cloud platform.

To some extent, datacenters can synchronize their energy consumption with their own production (thanks to load shifting and energy storage systems) or capitalize on collective self-consumption initiatives. The goal of all these efforts is to consume the least possible amount of brown energy and make profit. Cooling systems also contribute significantly to energy consumption (and therefore to carbon footprint) and water consumption in datacenters. Interestingly, they are flexible entities. Hence, some studies take into account these variations in the cooling system to optimize datacenter operation.

4.4.3 Operating sustainable datacenters through variations in the cooling system

According to the IEA, "*electricity demand in datacenters is mainly from two processes, with computing accounting for 40% of electricity demand of a datacenter. Cooling requirements to achieve stable processing efficiency similarly makes up about another 40%. The remaining 20% comes from other associated IT equipment*" [55]. Moreover, as discussed in Section 4.1.3, datacenters have reached a worrisome level of water consumption mainly driven by cooling and electricity generation. On another hand, recent researches observe that the cooling system energy efficiency [20, 60] and water effectiveness [46, 60] are in fact dynamic and influenced by external weather conditions. This has led to the development of job scheduling and resource allocation policies that take into account the real-time local conditions (weather, water availability, electrical grid mix, water requirement for cooling and generating energy, etc.) to schedule or balance load in distributed infrastructures, while respecting the service level agreements (SLAs) set with users.

For instance, based on a single prototype datacenter, the authors of [68] conduct a study on reducing electricity cost and environmental impact. Their approach combines several aspects for tasks planning: using renewable energy sources, adjusting energy use based on price fluctuations, and dynamically adjusting the cooling system (that could switch between mechanical cooling and free cooling using outside air). The approach first predicts both the availability of renewable energy and the demand for IT resources. With these predictions, it schedules arriving tasks and allocates resources within the datacenter to take advantage of available renewable energy and to optimize cooling efficiency. Simulating the approach on real-world data from datacenters, the authors achieve a reduction of both power costs and non-renewable energy usage by up to 60%, while meeting SLAs.

The authors of [56] propose a batch job scheduling algorithm, called minimization of WAter, Carbon and Electricity cost (WACE) that exploits the inherent spatio-temporal diversity of water efficiency across geographically distributed datacenters. WACE dynamically adjusts geographic load balancing and resource provisioning to minimize the water consumption along with carbon emissions and electricity cost, while satisfying average delay performance requirement. The algorithm also employs resource consolidation and DVFS. Simulation results on real delay-tolerant traces show that the method reduces the total cost by 20% and the water consumption by 25%, while satisfying the average delay constraint. Then, the authors extend WACE to jointly schedule delay-tolerant batch jobs and delay-sensitive interactive jobs.

In [39], the authors conduct a study on diverse cooling system designs in the perspective of jointly mitigating the wasted water for both cooling and electricity production, the energy consumed for cooling, the associated carbon impact, and the costs linked with cooling. They first compare dry (i.e., without water loss) and evaporative (i.e., relying on water evaporation) cooling modes. While the first mode is less energy-efficient, it does not consume water. Then, they quantify the water savings enabled by dry cooling and the converse overhead of energy, carbon and cost that it represents. As a solution, they propose to operate the equipment slightly out of their reliable specification to improve the energy efficiency of the dry cooling, but this practice does not offset the energy overhead between dry and evaporative cooling. For the cost, they propose employing the redundancy equipment more actively. That solution offsets the cost overhead of dry cooling and even proves to be more beneficial than evaporative cooling, although it is important to note that it may come at the cost of an increased risk of failure. Finally, they propose two optimization approaches allowing to dynamically make the compromise between water, energy, carbon and cost. These approaches are based on hybrid cooling systems alternating between dry and evaporative modes.

There are numerous contributions considering various optimization criteria with different prioritization among these criteria. Table 4.2 summarizes the literature work optimizing the computing and electrical infrastructures as well as their management.

4.5 Conclusion

In this chapter, we have examined the evolution of computing infrastructures, from the centralized ones (Cloud-type) to the decentralized ones (Edge-types), as well as their impact on the environment. We have also presented the

Table 4.2: Adapting computation in response to the electrical system and/or cooling system.

References	Considered criteria					
	Meeting QoS requirement	Energy saving	Water saving	Reducing CO_2	Minimizing costs	Flexibility used
[44, 110]	✓	×	×	✓	✓	Renewables management & task scheduling
[48, 101]	✓	×	×	✓	✓	AI-based Renewables management & task scheduling
[63]	✓	×	×	✓	✓	Virtual machine (VM) migration
[17]	✓	✓	×	✓	✓	Renewables aware scheduling & VM migration & consolidation
[92]	✓	×	×	✓	✓	Spatio-temporal load shifting
[40, 65, 66]	✓	×	×	✓	✓	Renewables & storage & task scheduling
[78]	✓	×	×	✓	×	Autonomy & renewables & storage & power profiling
[47, 64]	✓	×	×	✓	✓	Autonomy & renewables & storage
[20]	×	×	×	✓	✓	Flexible PUE analysis
[60]	×	✓	✓	✓	✓	Water-energy trade-off analysis
[46]	×	×	✓	✓	×	Water consumption modeling & dataset
[68]	✓	✓	✓	✓	✓	Renewable & cooling aware task scheduling
[56]	✓	✓	✓	✓	✓	DVFS & consolidation & Renewable + cooling aware spatio-temporal load shifting
[39]	✓	✓	✓	✓	✓	Out-of-bounds operation & redundancy reduction

basic components of these infrastructures' electrical and cooling systems, along with ways of greening them. We have then reviewed the literature on flexibilities used to improve the service and/or reduce the environmental impact of these infrastructures. These flexibilities include infrastructure and electrical system design, infrastructure operation through hardware features and efficient software, and the dynamics of the cooling and electrical systems. These many levels of flexibility in design and operation provide a fertile ground for *co-optimization* across abstractions for improved sustainability.

4.6 References

[1] M. Abdelshkour. IoT, from Cloud to Fog Computing. Cisco blogs, `https://blogs.cisco.com/perspectives/iot-from-cloud-to-fog-computing`, 2015. Accessed:17/06/2024.

[2] E. Ahvar, A.-C. Orgerie, and A. Lebre. Estimating energy consumption of cloud, fog, and edge computing infrastructures. *IEEE Transactions on Sustainable Computing*, 7(2):277–288, 2019.

[3] S. Al-Sarawi, M. Anbar, R. Abdullah, and A. B. Al Hawari. Internet of Things Market Analysis Forecasts, 2020–2030. In *World Conference on Smart Trends in Systems, Security and Sustainability (WorldS4)*, pages 449–453, 2020.

[4] M. P. Alves, F. C. Delicato, I. L. Santos, and P. F. Pires. LW-CoEdge: a lightweight virtualization model and collaboration process for edge computing. *World Wide Web*, 23:1127–1175, 2020.

[5] Amazon. Reducing water usage in AWS data centers. `https://www.aboutamazon.com/news/aws/reducing-water-usage-in-aws-data-centers`, 2020. Accessed: 2023-10-23.

[6] G. Ananthanarayanan, P. Bahl, P. Bodík, K. Chintalapudi, M. Philipose, L. Ravindranath, and S. Sinha. Real-Time Video Analytics: The Killer App for Edge Computing. *Computer*, 50(10):58–67, 2017.

[7] G. Antoniou, H. Volos, D. B. Bartolini, T. Rollet, Y. Sazeides, and J. H. Yahya. AgilePkgC: An Agile System Idle State Architecture for Energy Proportional Datacenter Servers. In *IEEE/ACM International Symposium on Microarchitecture (MICRO)*, pages 851–867, 2022.

[8] A. Aqil, A. O. F. Atya, S. V. Krishnamurthy, and G. Papageorgiou. Streaming Lower Quality Video over LTE: How Much Energy Can You Save? In *IEEE International Conference on Network Protocols (ICNP)*, pages 156–167, 2015.

[9] A. Asghari and M. K. Sohrabi. Server placement in mobile cloud computing: a comprehensive survey for edge computing, fog computing and cloudlet. *Computer Science Review*, 51, 2024.

[10] M. S. Aslanpour, S. S. Gill, and A. N. Toosi. Performance evaluation metrics for cloud, fog and edge computing: A review, taxonomy, benchmarks and standards for future research. *Internet of Things*, 12, 2020.

[11] M. Auvil. Reducing water usage in cooling systems to boost sustainability. DataCenter Knowlege, `https://www.datacenterknowledge.com/industry-perspectives/reducing-water-usage-cooling-systems-boost-sustainability`, 2022. Accessed: 2023-10-23.

[12] H. Badri, T. Bahreini, D. Grosu, and K. Yang. Energy-Aware Application Placement in Mobile Edge Computing: A Stochastic Optimization Approach. *IEEE Transactions on Parallel and Distributed Systems*, 31(4):909–922, 2020.

[13] L. A. Barroso, U. Hölzle, and P. Ranganathan. Data Center Basics: Building, Power, and Cooling. In *The Datacenter as a Computer*, pages 75–98. Springer, Cham, 2019.

[14] N. Bashir, N. Deng, K. Rzadca, D. Irwin, S. Kodak, and R. Jnagal. Take it to the limit: peak prediction-driven resource overcommitment in datacenters. In *European Conference on Computer Systems*, pages 556–573, 2021.

[15] A. Beloglazov, J. Abawajy, and R. Buyya. Energy-aware resource allocation heuristics for efficient management of data centers for cloud computing. *Future generation computer systems*, 28(5):755–768, 2012.

[16] K. Brandt. Our commitment to water stewardship. Google blog, https://blog.google/outreach-initiatives/sustainability/replenishing-water/, 2021. Accessed: 2023-10-29.

[17] B. Camus, A. Blavette, F. Dufossé, and A.-C. Orgerie. Harnessing the geographical flexibility of distributed computing clouds for cooperative self-consumption. In *IEEE PES Innovative Smart Grid Technologies (ISGT) Europe*, 2018.

[18] Z. Cao, X. Zhou, H. Hu, Z. Wang, and Y. Wen. Toward a systematic survey for carbon neutral data centers. *IEEE Communications Surveys & Tutorials*, 24(2):895–936, 2022.

[19] L. Chen and A. P. Wemhoff. Characterizing Data Center Cooling System Water Stress in the United States. *ASHRAE Annual Conference papers*, pages 1–9, 2022.

[20] A. A. Chien, C. Zhang, and L. Lin. Beyond PUE: Flexible datacenters empowering the cloud to decarbonize. *USENIX Hot Carbon*, 2022.

[21] J. E. Contreras-Ocaña, A. Singh, Y. Bésanger, and F. Wurtz. Integrated Planning of a Solar/Storage Collective. *IEEE Transactions on Smart Grid*, 12(1):215–226, 2021.

[22] C. Courageux-Sudan. *End-to-end simulation of the energy consumption of Fog infrastructures and their applications.* PhD thesis, Université de Rennes, 2023.

[23] H. David, E. Gorbatov, U. R. Hanebutte, R. Khanna, and C. Le. RAPL: Memory power estimation and capping. In *ACM/IEEE international symposium on Low power electronics and design*, pages 189–194, 2010.

[24] P. Devarasetty and S. Reddy. Genetic algorithm for quality of service based resource allocation in cloud computing. *Evolutionary Intelligence*, 14:381–387, 2021.

[25] ETSI. Mobile-edge computing - introductory technical White Paper. Technical report, ETSI, https://portal.etsi.org/Portals/0/TBpages/MEC/Docs/Mobile-edge_Computing_-_Introductory_Technical_White_Paper_V1%2018-09-14.pdf, 2014. Accessed:17/06/2024.

[26] European Commission. Energy Communities. https://energy.ec.europa.eu/topics/markets-and-consumers/energy-consumers-and-prosumers/energy-communities_en. Accessed: 2024-07-23.

[27] European Commission. Questions and Answers: EU action plan on digitalising the energy system. https://ec.europa.eu/commission/presscorner/detail/en/qanda_22_6229, 2022.

[28] G. Fieni, D. R. Acero, P. Rust, and R. Rouvoy. PowerAPI: A Python framework for building software-defined power meters. *Journal of Open Source Software*, 9(98):6670, 2024.

[29] Fitzgerald, David. Data Centers and Hidden Water Use. Wall Street Journal, www.wsj.com/articles/data-centers-1435168386, 2015. Accessed: 2024-06-18.

[30] C. Freitag, M. Berners-Lee, K. Widdicks, B. Knowles, G. S. Blair, and A. Friday. The real climate and transformative impact of ICT: A critique of estimates, trends, and regulations. *Patterns*, 2(9), 2021.

[31] French Government. Arrêté du 21 novembre 2019 fixant le critère de proximité géographique de l'autoconsommation collective étendue. French legislation on the Collective Self Consumption, https://www.legifrance.gouv.fr/loda/id/JORFTEXT000039417566/, 2020.

[32] D. Frieden, A. Tuerk, C. Neumann, S. d'Herbemont, and J. Roberts. Collective self-consumption and energy communities: Trends and challenges in the transposition of the EU framework. Working paper of the COMPILE project, https://www.rescoop.eu/uploads/rescoop/downloads/Collective-self-consumption-and-energy-communities.-Trends-and-challenges-in-the-transposition-of-the-EU-framework.pdf, 2020.

[33] D. Frieden, A. Tuerk, J. Roberts, S. D'Herbemont, A. F. Gubina, and B. Komel. Overview of emerging regulatory frameworks on collective self-consumption and energy communities in Europe. In *International Conference on the European Energy Market (EEM)*, pages 1–6, 2019.

[34] GHG Protocol. A Corporate Accounting and Reporting Standard. https://ghgprotocol.org/sites/default/files/standards/ghg-protocol-revised.pdf, 2004. Accessed: 2025-12-01.

[35] D. Gluck and R. Mazzoli. About Microsoft Datacenters. Microsoft, https://learn.microsoft.com/en-us/compliance/assurance/assurance-datacenter-architecture-infrastructure, 2024. Accessed: 2024-06-14.

[36] W. E. Gnibga, A. Blavette, and A.-C. Orgerie. Latency, Energy and Carbon Aware Collaborative Resource Allocation with Consolidation and QoS Degradation Strategies in Edge Computing. In *IEEE International Conference on Parallel and Distributed Systems (ICPADS)*, pages 2630–2639, 2023.

[37] W. E. Gnibga, A. Blavette, and A.-C. Orgerie. Renewable Energy in Data Centers: the Dilemma of Electrical Grid Dependency and Autonomy Costs. *IEEE Transactions on Sustainable Computing*, 9(3):315–328, 2023.

[38] W. E. Gnibga, A. Blavette, and A.-C. Orgerie. Energy-related Impact of Redefining Self-consumption for Distributed Edge Datacenters. In *International Green and Sustainable Computing Conference (IGSC)*, pages 7–13, 2024.

[39] W. E. Gnibga, A. A. Chien, A. Blavette, and A.-C. Orgerie. FlexCoolDC: Datacenter Cooling Flexibility for Harmonizing Water, Energy, Carbon, and Cost Trade-offs. In *ACM International Conference on Future and Sustainable Energy Systems (e-Energy)*, pages 108–122, 2024.

[40] Í. Goiri, W. Katsak, K. Le, T. D. Nguyen, and R. Bianchini. Parasol and greenswitch: Managing datacenters powered by renewable energy. *ACM SIGPLAN Notices*, 48(4):51–64, 2013.

[41] Í. Goiri, K. Le, J. Guitart, J. Torres, and R. Bianchini. Intelligent placement of datacenters for internet services. In *International Conference on Distributed Computing Systems*, pages 131–142, 2011.

[42] Google. Google Distributed Cloud. `https://cloud.google.com/distributed-cloud`, 2024. Accessed: 2024-06-14.

[43] A. Gordon, M. Hines, D. Da Silva, M. Ben-Yehuda, M. Silva, and G. Lizarraga. Ginkgo: Automated, application-driven memory overcommitment for cloud computing. In *Runtime Environments/Systems, Layering, and Virtualized Environments Workshop (ASPLOS RESoLVE workshop)*, 2011.

[44] L. Grange, G. Da Costa, and P. Stolf. Green IT scheduling for data center powered with renewable energy. *Future Generation Computer Systems*, 86:99–120, 2018.

[45] A. Griffiths, A. Morsman, and P. Veitch. Understanding the Performance and Power Saving Tradeoffs of Server Sleep States. In *IEEE International Conference on Cloud Networking (CloudNet)*, pages 94–102, 2023.

[46] P. S. Gupta, M. R. Hossen, P. Li, S. Ren, and M. A. Islam. A dataset for research on water sustainability. In *ACM International Conference on Future and Sustainable Energy Systems*, pages 442–446, 2024.

[47] M. Haddad, G. Da Costa, J.-M. Nicod, M.-C. Péra, J.-M. Pierson, V. Rehn-Sonigo, P. Stolf, and C. Varnier. Combined IT and power supply infrastructure sizing for standalone green data centers. *Sustainable Computing: Informatics and Systems*, 30, 2021.

[48] H. He, H. Shen, Q. Hao, and H. Tian. Online delay-guaranteed workload scheduling to minimize power cost in cloud data centers using renewable energy. *Journal of Parallel and Distributed Computing*, 159:51–64, 2022.

[49] Q. He, G. Cui, X. Zhang, F. Chen, S. Deng, H. Jin, Y. Li, and Y. Yang. A game-theoretical approach for user allocation in edge computing environment. *IEEE Transactions on Parallel and Distributed Systems*, 31(3):515–529, 2019.

[50] L. Helali and M. N. Omri. A survey of data center consolidation in cloud computing systems. *Computer Science Review*, 39, 2021.

[51] P. Hu, S. Dhelim, H. Ning, and T. Qiu. Survey on fog computing: architecture, key technologies, applications and open issues. *Journal of network and computer applications*, 98:27–42, 2017.

[52] Hubblo. Scaphandre. `https://github.com/hubblo-org/scaphandre`, 2023.

[53] R. Inta, D. J. Bowman, and S. M. Scott. The "Chimera": An Off-The-Shelf CPU/GPGPU/FPGA Hybrid Computing Platform. *International Journal of Reconfigurable Computing*, 2012(1):241439, 2012.

[54] International Energy Agency. Data Centres and Data Transmission Networks. `https://www.iea.org/energy-system/buildings/data-centres-and-data-transmission-networks`, 2023.

[55] International Energy Agency. Electricity 2024 analysis and forecast to 2026. `https://www.iea.org/reports/electricity-2024`, 2024.

[56] M. A. Islam, K. Ahmed, H. Xu, N. H. Tran, G. Quan, and S. Ren. Exploiting Spatio-Temporal Diversity for Water Saving in Geo-Distributed Data Centers. *IEEE Transactions on Cloud Computing*, 6(3):734–746, 2018.

[57] M. Jay, V. Ostapenco, L. Lefèvre, D. Trystram, A.-C. Orgerie, and B. Fichel. An experimental comparison of software-based power meters: focus on CPU and GPU. In *IEEE/ACM International Symposium on Cluster, Cloud and Internet Computing (CCGrid)*, pages 106–118, 2023.

[58] C. Jiang, T. Fan, H. Gao, W. Shi, L. Liu, C. Cérin, and J. Wan. Energy aware edge computing: A survey. *Computer Communications*, 151:556–580, 2020.

[59] K. Kaffes, D. Sbirlea, Y. Lin, D. Lo, and C. Kozyrakis. Leveraging application classes to save power in highly-utilized data centers. In *ACM Symposium on Cloud Computing*, pages 134–149, 2020.

[60] L. Karimi, L. Yacuel, J. Degraft-Johnson, J. Ashby, M. Green, M. Renner, A. Bergman, R. Norwood, and K. L. Hickenbottom. Water-energy tradeoffs in data centers: A case study in hot-arid climates. *Resources, Conservation and Recycling*, 181:1–10, 2022.

[61] M. I. Khaleel, M. Safran, S. Alfarhood, and D. Gupta. Combinatorial metaheuristic methods to optimize the scheduling of scientific workflows in green DVFS-enabled edge-cloud computing. *Alexandria Engineering Journal*, 86:458–470, 2024.

[62] W. Z. Khan, E. Ahmed, S. Hakak, I. Yaqoob, and A. Ahmed. Edge computing: A survey. *Future Generation Computer Systems*, 97:219–235, 2019.

[63] A. Khosravi, A. Nadjaran Toosi, and R. Buyya. Online virtual machine migration for renewable energy usage maximization in geographically distributed cloud data centers. *Concurrency and Computation: Practice and Experience*, 29, 2017.

[64] N. Lazaar, M. Barakat, M. Hafiane, J. Sabor, and H. Gualous. Modeling and control of a hydrogen-based green data center. *Electric Power Systems Research*, 199, 2021.

[65] H. Li, M. Alaküla, and H. Gualous. A Forecasting Based Hierarchical Energy Management for Sustainable Data Centers. In *Annual Conference of the IEEE Industrial Electronics Society (IECON)*, pages 01–06, 2023.

[66] Y. Li, A.-C. Orgerie, and J.-M. Menaud. Balancing the use of batteries and opportunistic scheduling policies for maximizing renewable energy consumption in a cloud data center. In *Euromicro International Conference on Parallel, Distributed and Network-based Processing (PDP)*, pages 408–415, 2017.

[67] J. Liu, P. Yang, and C. Chen. Intelligent energy-efficient scheduling with ant colony techniques for heterogeneous edge computing. *Journal of Parallel and Distributed Computing*, 172:84–96, 2023.

[68] Z. Liu, Y. Chen, C. Bash, A. Wierman, D. Gmach, Z. Wang, M. Marwah, and C. Hyser. Renewable and cooling aware workload management for sustainable data centers. In *ACM SIGMETRICS/PERFORMANCE joint international conference on Measurement and Modeling of Computer Systems*, pages 175–186, 2012.

[69] G. Lovász, F. Niedermeier, and H. De Meer. Performance tradeoffs of energy-aware virtual machine consolidation. *Cluster Computing*, 16:481–496, 2013.

[70] Q. Luo, S. Hu, C. Li, G. Li, and W. Shi. Resource scheduling in edge computing: A survey. *IEEE Communications Surveys & Tutorials*, 23(4):2131–2165, 2021.

[71] M. Malik, H. Ghasemzadeh, T. Mohsenin, R. Cammarota, L. Zhao, A. Sasan, H. Homayoun, and S. Rafatirad. Ecost: Energy-efficient co-locating and self-tuning mapreduce applications. In *International Conference on Parallel Processing*, pages 1–11, 2019.

[72] I. Mohiuddin and A. Almogren. Workload aware VM consolidation method in edge/cloud computing for IoT applications. *Journal of Parallel and Distributed Computing*, 123:204–214, 2019.

[73] R. Morabito, V. Cozzolino, A. Y. Ding, N. Beijar, and J. Ott. Consolidate IoT edge computing with lightweight virtualization. *IEEE network*, 32(1):102–111, 2018.

[74] R. Muralidhar, R. Borovica-Gajic, and R. Buyya. Energy efficient computing systems: Architectures, abstractions and modeling to techniques and standards. *ACM Computing Surveys (CSUR)*, 54(11s):1–37, 2022.

[75] S. Mustafa, K. Bilal, S. A. Madani, N. Tziritas, S. U. Khan, and L. T. Yang. Performance Evaluation of Energy-Aware Best Fit Decreasing Algorithms for Cloud Environments. In *IEEE International Conference on Data Science and Data Intensive Systems*, pages 464–469, 2015.

[76] D. Mytton. Data centre water consumption. *npj Clean Water*, 4(1):11, 2021.

[77] G. Perin, M. Berno, T. Erseghe, and M. Rossi. Towards sustainable edge computing through renewable energy resources and online, distributed and predictive scheduling. *IEEE Transactions on Network and Service Management*, 19(1):306–321, 2021.

[78] J.-M. Pierson et al. Datazero: Datacenter With Zero Emission and Robust Management Using Renewable Energy. *IEEE Access*, 7:209–230, 2019.

[79] Q. Qu, R. Xu, S. Y. Nikouei, and Y. Chen. An Experimental Study on Microservices based Edge Computing Platforms. In *IEEE INFOCOM 2020 - IEEE Conf. on Computer Communications Workshops*, pages 836–841, 2020.

[80] A. Radovanović, R. Koningstein, I. Schneider, B. Chen, A. Duarte, B. Roy, D. Xiao, M. Haridasan, P. Hung, N. Care, S. Talukdar,

E. Mullen, K. Smith, M. Cottman, and W. Cirne. Carbon-Aware Computing for Datacenters. *IEEE Transactions on Power Systems*, 38(2):1270–1280, 2023.

[81] V. D. Reddy, B. Setz, G. S. V. R. K. Rao, G. R. Gangadharan, and M. Aiello. Metrics for Sustainable Data Centers. *IEEE Transactions on Sustainable Computing*, 2(3):290–303, 2017.

[82] D. Rosendo, A. Costan, P. Valduriez, and G. Antoniu. Distributed intelligence on the Edge-to-Cloud Continuum: A systematic literature review. *Journal of Parallel and Distributed Computing*, 166:71–94, 2022.

[83] S. Sarangi, E. D. McAfee, D. G. Damm, and J. Gullbrand. Single-Phase Immersion Cooling Performance in Intel Servers with Immersion Influenced Heatsink Design. In *Semiconductor Thermal Measurement, Modeling & Management Symposium*, pages 1–5, 2022.

[84] T. Scheibe. Cisco Distributed Data Centers Are Ready for 5G. CISCO, `https://blogs.cisco.com/datacenter/cisco-distributed-data-centers-are-ready-for-5g`, 2020. Accessed: 2024-06-14.

[85] D. Selvapandian and R. Santosh. A hybrid optimized resource allocation model for multi-cloud environment using bat and particle swarm optimization algorithms. *Computer Assisted Methods in Engineering and Science*, 29(1–2):87–103, 2022.

[86] T. Sen and H. Shen. Machine learning based timeliness-guaranteed and energy-efficient task assignment in edge computing systems. In *IEEE International Conference on Fog and Edge Computing (ICFEC)*, pages 1–10, 2019.

[87] T. Shi, H. Ma, and G. Chen. Energy-aware container consolidation based on PSO in cloud data centers. In *IEEE Congress on Evolutionary Computation (CEC)*, pages 1–8, 2018.

[88] M. A. B. Siddik, A. Shehabi, and L. Marston. The environmental footprint of data centers in the United States. *Environmental Research Letters*, 16(6):064017, 2021.

[89] L. Silva-Llanca, C. Ponce, E. Bermúdez, D. Martínez, A. J. Díaz, and F. Aguirre. Improving energy and water consumption of a data center via air free-cooling economization: The effect weather on its performance. *Energy Conversion and Management*, 292:1–19, 2023.

[90] B. Smith. Microsoft will replenish more water than it consumes by 2030. Microsoft blog, `https://blogs.microsoft.com/blog/2020/09/21/microsoft-will-replenish-more-water-than-it-consumes-by-2030/`, 2020. Accessed: 2023-10-29.

[91] M. Stephant, K. Hassam-Ouari, D. Abbes, A. Labrunie, and B. Robyns. A survey on energy management and blockchain for collective self-consumption. In *International Conference on Systems and Control (ICSC)*, pages 237–243, 2018.

[92] T. Sukprasert, A. Souza, N. Bashir, D. Irwin, and P. Shenoy. On the Limitations of Carbon-Aware Temporal and Spatial Workload Shifting in the Cloud. In *European Conference on Computer Systems*, pages 924–941, 2024.

[93] Z. Sun, G. Sun, Y. Liu, J. Wang, and D. Cao. Bargain-Match: A game theoretical approach for resource allocation and task offloading in vehicular edge computing networks. *IEEE Transactions on Mobile Computing*, 23(2):1655–1673, 2023.

[94] Z. Tang, W. Jia, X. Zhou, W. Yang, and Y. You. Representation and reinforcement learning for task scheduling in edge computing. *IEEE Transactions on Big Data*, 8(3):795–808, 2020.

[95] Z. Tao, Q. Xia, Z. Hao, C. Li, L. Ma, S. Yi, and Q. Li. A Survey of Virtual Machine Management in Edge Computing. *Proceedings of the IEEE*, 107(8):1482–1499, 2019.

[96] Z. Thomas. Could Underwater Data Centers Make Cloud Computing Greener? Wall Street Journal, `https://www.wsj.com/podcasts/tech-news-briefing/could-underwater-data-centers-make-cloud-computing-greener/7d3d778e-dae5-4a23-b6fd-9caacd268821`, 2022.

[97] J. Treibig, G. Hager, and G. Wellein. Likwid: A lightweight performance-oriented tool suite for x86 multicore environments. In *2010 39th international conference on parallel processing workshops*, pages 207–216. IEEE, 2010.

[98] US Environmental Protection Agency. Overview of Greenhouse Gases. `https://www.epa.gov/ghgemissions/overview-greenhouse-gases`, 2022. Accessed: 26/06/2024.

[99] US Environmental Protection Agency. Water Conservation at EPA. `https://web.archive.org/web/20230619145728/https://www.epa.gov/greeningepa/water-conservation-epa`, 2023. Accessed: 2023-06-19.

[100] B. Varghese, N. Wang, S. Barbhuiya, P. Kilpatrick, and D. S. Nikolopoulos. Challenges and opportunities in edge computing. In *IEEE International Conference on Smart Cloud (SmartCloud)*, pages 20–26, 2016.

[101] V. Venkataswamy, J. Grigsby, A. Grimshaw, and Y. Qi. RARE: Renewable Energy Aware Resource Management in Datacenters. In *Workshop on Job Scheduling Strategies for Parallel Processing*, 2022.

[102] B. Whitehead, R. Tozer, D. Andrews, and S. Flucker. Energy and water environmental trade-offs of data center cooling technologies. *ASHRAE Annual Conference papers*, pages 1–9, 2017.

[103] W. Xia and L. Shen. Joint resource allocation at edge cloud based on ant colony optimization and genetic algorithm. *Wireless Personal Communications*, 117(2):355–386, 2021.

[104] H. Yin, X. Zhang, H. H. Liu, Y. Luo, C. Tian, S. Zhao, and F. Li. Edge provisioning with flexible server placement. *IEEE Transactions on Parallel and Distributed Systems*, 28(4):1031–1045, 2016.

[105] M. Zakarya. Energy, performance and cost efficient datacenters: A survey. *Renewable and Sustainable Energy Reviews*, 94:363–385, 2018.

[106] H. Zhang and H. Hoffmann. PoDD: power-capping dependent distributed applications. In *International Conference for High Performance Computing, Networking, Storage and Analysis*, pages 1–23, 2019.

[107] J. Zhang, H. Huang, and X. Wang. Resource provision algorithms in cloud computing: A survey. *Journal of network and computer applications*, 64:23–42, 2016.

[108] Q. Zhang, Z. Meng, X. Hong, Y. Zhan, J. Liu, J. Dong, T. Bai, J. Niu, and M. J. Deen. A survey on data center cooling systems: Technology, power consumption modeling and control strategy optimization. *Journal of Systems Architecture*, 119:1–17, 2021.

[109] Z. Zhu, W. Zhang, V. Chaturvedi, and A. K. Singh. Energy minimization for multicore platforms through DVFS and VR phase scaling with comprehensive convex model. *IEEE Transactions on Computer-Aided Design of Integrated Circuits and Systems*, 39(3):686–699, 2019.

[110] Íñigo Goiri, M. E. Haque, K. Le, R. Beauchea, T. D. Nguyen, J. Guitart, J. Torres, and R. Bianchini. Matching renewable energy supply and demand in green datacenters. *Ad Hoc Networks*, 25:520–534, 2015.

Chapter 5

A Comprehensive Blueprint for Sustainable Computing

Pierre Segonne

This chapter examines the critical intersection between datacenter operations and electricity grid sustainability, presenting a comprehensive analysis of how computing services can contribute to grid decarbonization. We first establish that the ICT sector's emissions, particularly from datacenters, are growing at an unprecedented rate that outpaces efficiency gains. Through an examination of electricity grid operations and carbon accounting methodologies, we demonstrate why common approaches to sustainable computing may have limited real-world impact on emissions reduction. The chapter critically evaluates carbon-aware computing strategies, revealing important limitations in their effectiveness when implemented at scale. We propose a novel four-step blueprint for sustainable computing that prioritizes actions based on their verified impact on emissions reduction: (1) reducing total energy consumption through efficient resource allocation, (2) locating compute resources in regions with abundant low-carbon electricity, (3) selecting providers committed to additional low-carbon electricity procurement, and (4) optimizing workload scheduling to match clean electricity availability. This framework provides organizations with clear, actionable guidance for real-world emissions reduction while supporting grid decarbonization efforts.

5.1 Introduction

For humanity to limit the effect of global warming to an average of 2° C above pre-industrial levels, greenhouse gas emissions must not exceed the carbon budget of 1150 $GtCO_2$ [15]. That directly implies that all industrial sectors must curb their emissions rapidly, yet the tech industry at large is at odds with this necessity. The information and communication technology (ICT) sector is estimated to be responsible for emissions covering 1.5 to 4% of total emissions [10] (that is roughly on par with the share of aviation [77]) and that number is expected to continue to increase at an astounding rate as digitization progresses and billions more people gain access to the Internet [9].

Emissions within ICT come from various sources: consumer devices, computers or smartphones; shared computing resources, such as datacenters; and networking infrastructure. Network and device emissions have remained relatively stable in the period 2020-2022, unlike those from datacenters at large, which fuel most of the sector's rising emissions, increasing by 46% during that same period [8].

While reliable ICT emissions data are difficult to source [8], the carbon footprint of large tech corporations is well documented and illustrates this trend. Emissions from Google, Meta and Tencent increased by 54% between 2020 and 2022, while emissions from Amazon Web Services (AWS), Azure and Google Cloud Platform (GCP) increased by 41% in the same range [8]. Although it is difficult to find the exact share, sustainability reports from these firms mention that the vast majority of the emissions accounted for result from the generation of the electricity consumed to power their operations [3, 45]. To this day, tech companies draw power from electricity grids, which contain a mix of electricity from different sources [31], including polluting fossil-fuel based ones, and they have to account for emissions resulting from this usage.

Datacenters are further generally concentrated geographically; the Republic of Ireland, for example, has attracted numerous datacenters thanks to its low corporate tax, which results in an unprecedented situation at a country level. In 2022, datacenters in Ireland consumed 5.3 TWh of electricity, representing a staggering 17% of all electricity consumed. In 2026, that share is expected to **exceed a third of all electricity consumed** [88]. Denmark is following suit, with the consumption of datacenters expected to amount to just under 20% of all electricity consumption by 2026 [88]. The installation of datacenters can thus pose an additional challenge to the decarbonisation of electricity grids. In fact, total consumption, partially driven by the expansion of datacenters, has increased by approximately 13.6% between 2017 and 2023 in Ireland [32].

Meeting a growing electricity demand with carbon-free electricity creates additional levels of complexity for power producers and grid operators. Not only is it necessary to replace fossil fuel-based thermal power plants with low-carbon alternatives, such as photovoltaic solar panels and wind turbines, but extra capacity must also be installed to meet the growing demand. To alleviate this issue, the Irish Commission for Regulation of Utilities published in 2021 new requirements for the installation and operations of datacenters. Datacenters must not be installed in grid-constrained regions; they must be designed together with enough on-site dispatchable generation (e.g., a diesel generator) or storage (e.g., a battery system) to meet their consumption; they must, moreover, accept reducing their overall electricity consumption when requested by grid operators [88]. The Irish regulation could set a precedent for datacenters worldwide, which could be required to offer consumption piloting capabilities to grid operators for their installation to be approved, similarly to contracts already offered to large businesses and industries [38]. This possibility has potentially contributed to the scramble of some of the biggest tech companies to ensure a constant and controlled supply of power. For example, Meta has published a request for proposal (RFP) for the installation of 1-4 GW of nuclear power to meet their energy needs in datacenters [62] and has announced the installation of a gas power plant with a capacity of 2 262 MW on the site of its future Northeast Louisiana datacenter [39].

Datacenters' efficiency has improved significantly since their onset. Google, for example, reports that its average power usage effectiveness (PUE) decreased from 1.23 in 2008 to 1.09 in 2024 [46], meaning that for 1 MWh of electrical energy consumed in their datacenters, the share that was not used effectively to power computing resources decreased from 23% to 9%. Alas, the increase in datacenters' usage far outweighs the efficiency gains, as demonstrated by the whopping 63% increase of total electricity usage for datacenters observed between 2020 and 2022 [8]. The power dynamics are not expected to reverse with this stark example of the Jevons paradox or rebound effect. The International Energy Agency (IEA) predicts that due, among others, to the expanding use of artificial intelligence (AI) and cryptocurrencies, the total amount of electricity used by datacenters could more than double from 2024 to 2026, reaching 1000 TWh then, equivalent to the total electricity consumed by Japan [88], which is the 5th largest country by consumed electricity.

We have thus exposed that the ICT sector needs to become a positive actor in humanity's movement to decarbonise its energy supply. Within the global sector, datacenters and computing resources are the most critically at odds with this ambition. Their expansion is so rapid that they now hold the power to disrupt the operation and decarbonisation of electricity grids worldwide, and far outstrips efficiency gains achieved by the industry in the last 15 years. If this development does not slow down, the tech sector will need to find a way to curb its emissions while expanding its operations. The overwhelming majority of its emissions come from the generation of electricity that powers the computing resources of datacenters. It therefore becomes essential for computing service providers to identify ways for datacenters to accelerate the global decarbonisation of electricity grids. That implies finding solutions that leverage the consumption of electricity in datacenters to encourage the deployment of low carbon electricity sources. This will only be possible when these solutions are aware of the physical reality of the grids and involve grid operators. That will become even more relevant as other sectors electrify (e.g., heating, transportation) and compete for access to cheap and reliable electricity access. As such, in this chapter, we will adopt the electricity grid perspective to examine different proposed solutions, and discuss their respective advantages and pitfalls.

5.2 Origin of the Electricity That Powers Computing

"When a megawatthour is drawn from the grid, where does that energy come from?" is a critical question to answer to be able to understand how to reduce emissions from electricity usage.

Historically, answering that question would have been easy; in the early industrial era, electricity would have come from the neighbouring coal power plant (for example, the Deptford power station would remain the only power plant in London for years). With time, electricity grids became more complex, integrating more numerous generation technologies (the first nuclear power plant was connected in 1954 in Obninsk; the first offshore wind farm was connected in 1991 in Vindeby), growing more interconnected (the first subsea interconnector linking the French and British grids entered operations in 1986), which made answering that question close to impossible for anyone not working for a transmission system operator (TSO). In a grid, it is indeed impossible to select the origin of the electricity that one consumes. Physically, the electricity injected into the grid is mixed, and at consumption, one inevitably receives a mix of electricity coming from different sources [23].

Fortunately, the last decade has seen public data about the operation of electricity grids become readily available. For example, in the context of the implementation of free markets for gas and power within the European Union (EU), the European Network of Transmission System Operators for Electricity (ENTSO-E) was established in 2009, regrouping all European TSOs and mandating them to freely share data about electricity generation, and exchanges within the EU.

The opening of electricity data has allowed the development of initiatives such as Electricity Maps [31], which determines the origin of the electricity consumed globally, at an hourly resolution. Other providers of grid data exist, but with data generally limited to production and not consumption mixes such as GridStatus [49], Energy-Charts [37], Singularity [80], NESO [65], or WattTime [81]. For Electricity Maps, a set of parsing scripts, largely developed by an open-source community, allows the constant collection of the most up-to-date reported generation, and electricity exchanges. From this raw data, the execution of flowtracing [24, 90] solves the task of tracing back the origin of electricity worldwide. Under the assumption of proportional mixing and irreversibility of power flows, such an algorithm resolves the set of equations that arises from applying the law of energy conservation within each zone of the global electricity network, even in the case of circular flows. It is expressed as follows:

$$E_{consumed} = E_{produced} + E_{imported} + E_{discharged} - E_{exported} - E_{stored}.$$

It should be noted that the inclusion of electricity exchanges between grids is paramount to consider if one wants to accurately represent the origin of electricity. Differences of more than 40% in the final carbon intensity have been observed when the electricity mix is computed without [26].

Carbon intensity, measured in grams of CO_2e per kWh, is the most common measure of the carbon content of the electricity on the grid. It describes the amount of greenhouse gas emissions that can be attributed to the consumption of each unit of electrical energy [29, 65, 85].

Exploring the evolution of the origin of electricity in different regions of the world reveals something crucial. The carbon intensity of electricity, which represents how emissions-intensive consumed electricity is, varies drastically. For example, the grid operated by the California ISO (CISO) is notable for the large penetration of solar power, leading to significant variations in its carbon intensity between day and night, as shown in Figure 5.1. Other grids have access to low-carbon electricity continuously, such as the one operated by the City of Tacoma. In contrast, some other grids struggle to access significant amounts of low carbon resources, such as the Northwestern Energy grid, as can be seen in Figure 5.2.

That means that executing a job within a datacenter in California at the wrong time of the day could see assigned emissions vary by a factor greater than 2. Similarly, executing that same job within the Northwestern Energy balancing authority instead of the City of Tacoma balancing authority could see these emissions multiplied by 18.

5.3 Carbon-Aware Computing

Many computing jobs exhibit an inherent flexibility. If one wants to process a large dataset, whether the computation required occurs in China, South Africa, or Canada makes no difference. Special fees might apply to transfer the data across the world, and its moving across networks might introduce some latency, but the output will be identical. Similarly, many asynchronous tasks do not need to be executed at an exact time point. Whether the reprocessing of the base data powering an analytical engine occurs at 3:00am or 4:13am (during the night) will not affect the work of data analysts within a company.

At the intersection of computing's flexibility and electricity grids' varying carbon intensity lies the idea of carbon awareness, or carbon-aware computing. The intention is to adapt the use of computational resources in time and/or space to minimize the resulting emissions [13, 28, 29]. As the Green Software Foundation (GSF) describes it [13], several approaches complement each other under the umbrella of carbon awareness.

- **Demand shifting** describes the intention of modifying the time or location of execution of a given task based on the carbon intensity [50, 85]. Some tasks, such as large analytical queries or machine learning training,

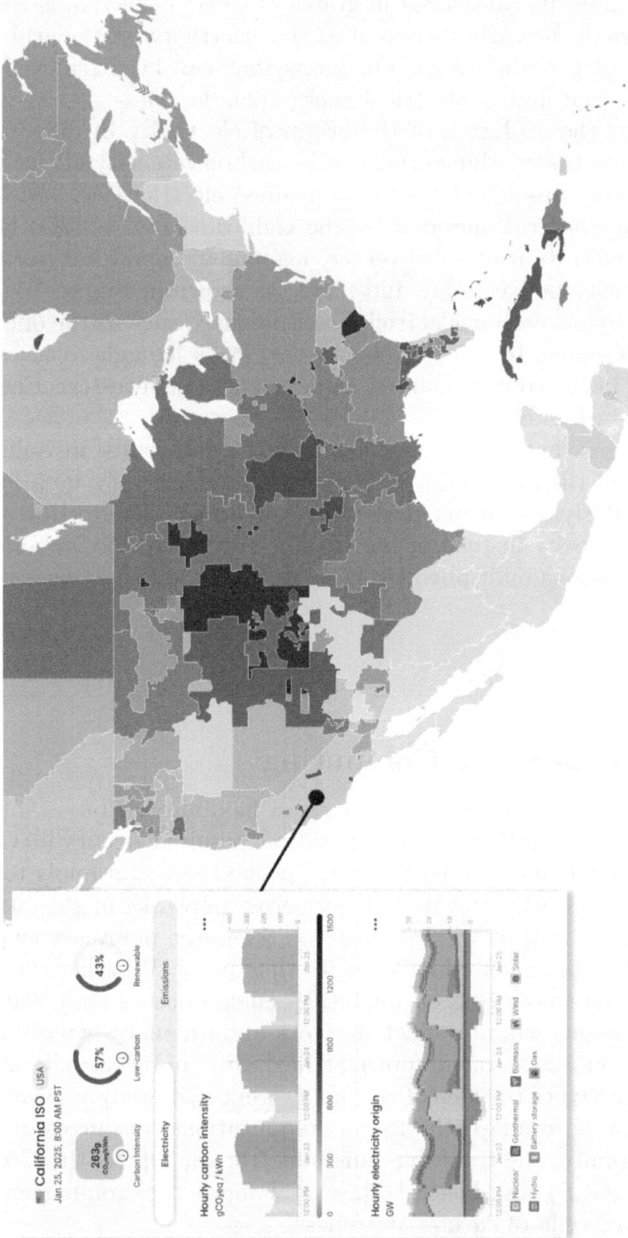

Figure 5.1: Carbon intensity and electrical mix for the California ISO. *Source: Electricity Maps* [31].

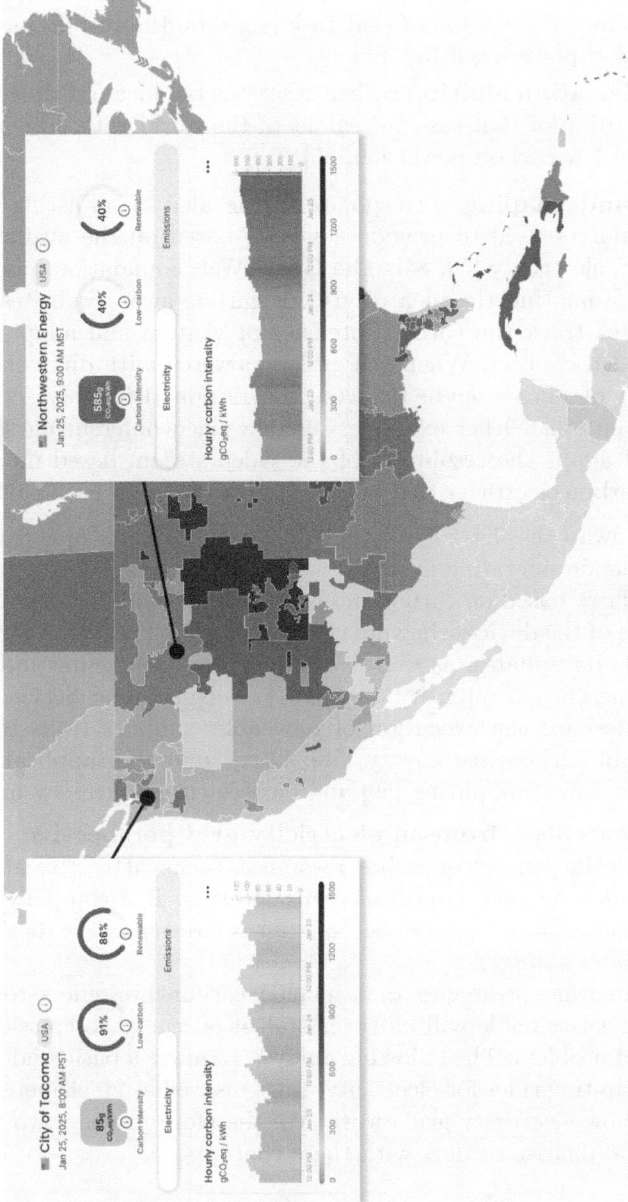

Figure 5.2: Carbon intensity comparison between two balancing authorities in the US. *Source: Electricity Maps* [31].

are particularly suited for this; tools such as the Carbon Aware SDK [84] are already available to help the development of carbon-aware services.

- **Time shifting** specifically describes the intention of adapting the time of execution of that task based on the carbon intensity of the grid powering it [50, 85].
- **Location shifting** rather describes the idea of displacing the execution of that task to regions of the world with a high availability of low carbon power [50, 85].

- **Demand shaping** corresponds to the idea of adjusting the amount of resources used to provide a service based on the availability of low carbon electricity [50, 85]. The Green Web Foundation has for example been promoting the idea of grid or carbon-aware websites [42]. These websites track the carbon intensity of visitors and adapt the amount of loaded content. When the grid is powered with dirty electricity, the quality of images can be reduced, and/or noncritical Javascript might be removed [56]. Other examples include videoconferencing software that would adapt the resolution of the video stream based on the share of low-carbon electricity that powers the computers of each participant [85].

Carbon awareness has already been integrated by multiple companies and services. Certain operating systems, such as Windows, control the execution of their updates based on carbon intensity [99]. Others, such as iOS, optimise the charging of the devices they are running on [85]. The XBox gaming console also controls its updates, as well as the lifecycle of its game and application catalog, using carbon intensity data [101]. Google Cloud Services is shifting in time and space the execution of moveable compute tasks to match the availability of carbon-free energy [58]. Other potential applications include carbon-aware edge computing [59] and flexible cryptocurrency mining [60].

Carbon awareness, from an electricity grid perspective

Although the concept of carbon awareness seems attractive at first glance, it leaves aside one very important consideration. If a company implements carbon-awareness into its services, *how can it reliably assess its effect on real-world carbon emissions?*

To ensure that strategies that include carbon awareness to make computing more sustainable will be successful, it is crucial to cross-examine this multifaceted problem. The following analysis requires a basic understanding of carbon accounting rules for electricity usage, as well as an elementary comprehension of how electricity grids work. We therefore provide two independent apartés to familiarise readers with these concepts.

Assigning Carbon Emissions from Electricity Usage

The Greenhouse Gas (GHG) Protocol Corporate Standards is the widely accepted framework for companies to account for greenhouse gas emissions in their inventories. It consists of a set of requirements and provides guidance for corporations to consistently account and disclose greenhouse gas emissions. Its use is ubiquitous in the latest regulations in Europe and the US. For example, the Corporate Sustainability Reporting Directive (CSRD) of the EU requires companies to disclose their impact on climate change using the European Sustainability Reporting Standards (ESRS), which requires the adoption of the GHG protocol rules. Similarly in the US, The Enhancement and Standardization of Climate-Related Disclosures for Investors Rule (SEC Rule) forces publicly traded companies to disclose parts of their greenhouse gas emissions as defined by the scope 1 and 2 of the GHG Protocol [63].

The GHG Protocol classifies companies' emissions into 3 scopes (see Figure 5.3):

- **Scope 1** emissions correspond to direct emissions from assets owned by the company. For example, if a tech company owns and operates backup diesel generators to ensure a constant supply of power to its datacenters, emissions from the exhaust gases of these generators would be accounted for under scope 1. The same would apply if cooling gases were leaked on site into the atmosphere.
- **Scope 2** emissions correspond to any indirect emissions resulting from the purchase of energy for its own use. If a company buys electricity produced by a coal power plant owned and operated by one of its suppliers, emissions resulting from the generation of electricity through the combustion of coal would be accounted for under the scope 2.
- **Scope 3** emissions correspond to the remaining indirect emissions that occur in the value chain of the company. That includes upstream emissions covering the goods and services the company uses and procures, as well as downstream emissions covering the usage and end-of-life treatment of the goods and services provided by the company.

Scope 2 - The dominant scope for electricity emissions

Scope 3 aside, companies selling computing services, such as datacenters and cloud operators, see most of their emissions fall under the scope 2 of the GHG Protocol. For example, the scope 1 emissions for Google accounted for 79.4 $ktCO_2e$, while the scope 2 covered 3.4 $MtCO_2e$ in 2023 [45]; the contrast would be even starker when looking exclusively at the cloud services part of Google. Strategies for making computing services greener should, therefore, focus on reducing scope 2 emissions.

Allocating grid emissions is more complex than what naive intuition might have us believe. As Electricity Maps' guide to *Understanding Electricity Scope 2 Attribution Rules* puts it [25]:

> There is unfortunately no scientifically proven way to determine the recipient of each unit of generated electricity, as there is unfortunately no way to "dye" a unit of generated electricity and observe where it ends up.

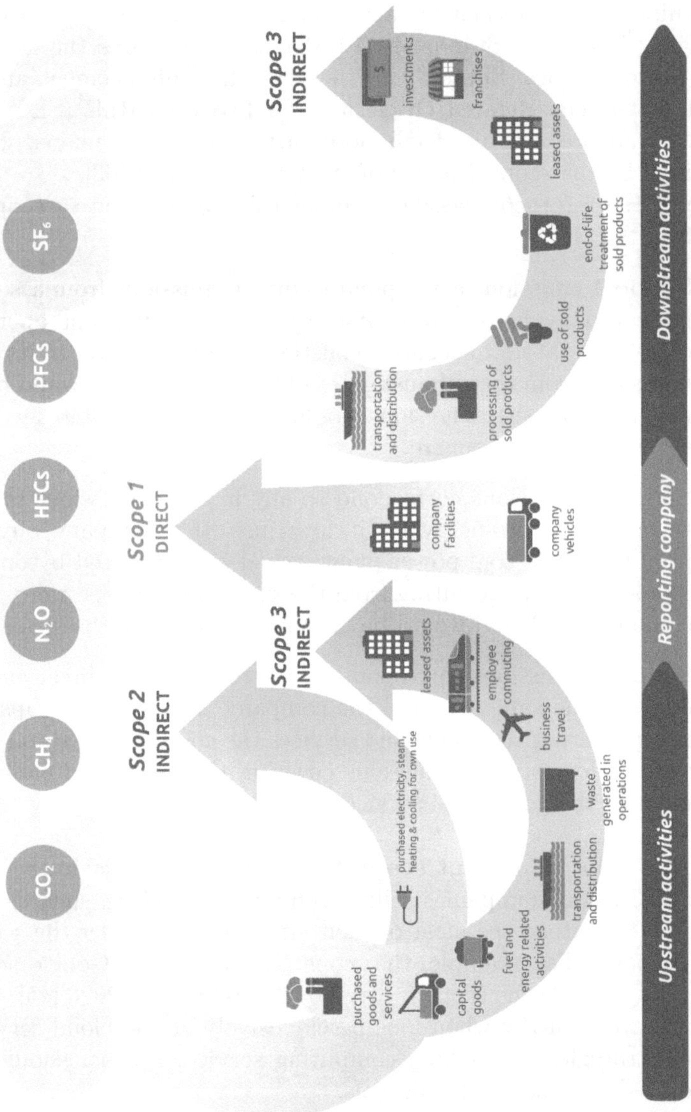

Figure 5.3: Overview of GHG Protocol scopes and emissions across the value chain. *Source: GHG Protocol's Technical Guidance for Calculating Scope 3 Emissions, page 6* [87].

This means that there is no unique and universally accepted way to directly attribute emissions to electricity consumers. Multiple accounting principles coexist under the scope 2 of the GHG Protocol, each with their advantages and drawbacks [73].

In general, accounting methods for scope 2 electricity emissions can be categorized under two main frameworks: location-based accounting and market-based accounting.

For **location-based accounting**, it is assumed that electricity consumers do not choose the origin of the electricity they consume. Whenever they draw power from the grid, they receive a mix of all the electricity injected into the grid, as would any other consumer. Although this attribution rule is very intuitive, highly verifiable, and credible, it provides little actionability for electricity consumers. The only way for them to reduce their scope 2 carbon footprint is either to wait for the whole grid to decarbonise, or to move all their operations to a cleaner grid.

Market-based accounting relies on the introduction of a market on which are sold certificates for the origin of the electricity injected into the grid. When a power producer generates 1 MWh of clean electricity, it will sell the electricity itself, as well as, and separately, an origin certificate matching that 1 MWh of clean electricity[1] [82]. The certificate will be usable by a consumer to claim that they have consumed exclusively 1 MWh of green energy. These certificates are called Energy Attribute Certificates (EACs) [94], Renewable Energy Certificates (RECs, only in the US) [94], or Guarantees of Origin (GOs, only in Europe) [7]. The market-based attribution rule assumes that consumers can choose the origin of their electricity [82]. Then, emissions from electricity consumption not covered by certificates are computed from the grid mix containing electricity that has not been claimed, called the residual mix [82]. Requirements for the usage of EACs vary, sometimes posing credibility issues with this method. Companies have, for example, been criticized [12] for claiming consumption of low carbon electricity from grids not physically connected to where consumption occurred or for covering electricity consumption at night with solar power. However, market-based accounting has the upside of providing greater actionability for consumers. Through the procurement of clean electricity contracts, companies can support financially clean power producers and take an active part in the decarbonisation of the grids. We note that certificates qualified as **additional** provide a stronger guarantee that this financial support will have a certain beneficial influence on the decarbonisation of the grids [27, 55, 76]. A certificate can be qualified as additional if it constituted a crucial contribution to the installation of new low-carbon electricity generation capacity.

This aparté on carbon accounting should provide the necessary knowledge to understand how committed computing services providers are to

[1]There are exceptions to this, such as bundled power purchase agreements (PPAs) where both the electricity and certificates are sold together.

participating in the decarbonisation of the grid. Most of their emissions will be reported under the scope 2 of the GHG protocol, and examining whether they report both location-based and market-based or only market-based emissions, as well as understanding what guidelines control their procurement of EACs, should shine a bright light on their commitment to leverage their power consumption to support the deployment of low-carbon electricity.

Balancing Electricity Grids

The goal behind the very existence of electrical grids is to provide electricity as efficiently, reliably and affordably as possible under a constantly changing demand. If demand was constant in time, electricity could easily be generated at a local level, within electricity islands, powered by a set of generators running at a constant power output. In reality, that is almost never the case. Demand can vary by more than 30% between its daily low, typically in the middle of the night, and its daily high, typically in the middle of the day [31].

Since the emergence of the first electricity grid, managing variation in demand has been a constant challenge. The balancing of the first grid built in Manhattan in 1882 by the Edison Illuminating company, which served up to 10,000 light bulbs, required constant attention. The generators were manned without interruption, and displayed visual signals to inform the crew of imbalances so that they could manually throttle their output to respond to people turning on or off their light bulbs [95].

Today, operators have better tools at their disposal to manage the constantly changing dynamics of their grids [64]. These tools, together with the immense scale of some modern electricity grids [93], make the balancing of electricity grids in some areas of the world much more reliable. For example, within the Synchronous grid of Continental Europe, which in 2009 integrated close to 670 GW of production capacity [93], a far cry from the 600 kW of the Pearl Street station, the last major blackout occurred in 2006 [92]. Due to this, grid operators have some flexibility to balance the demand and generation of electricity.

Sometimes, load balancing is oversimplified to the following: the amount of electricity consumed must match the amount of electricity produced at all times [98]. If that was the case, whenever demand increased, an already running generator should increase its power output, or a completely new generator should be turned on (vice-versa if the demand decreased). Imagine; if that was the case, how would grid operators manage to keep track of billions of events happening constantly, such as someone entering a new room and turning on a light bulb, and react accordingly?

Frequency regulation frees grid operators from needing absolute control over all inputs and outputs of the grid [57]. Within power grids operating on alternating current (AC), that is, all major grids currently in operation, the frequency of the current reflects the balance of demand and generation. If

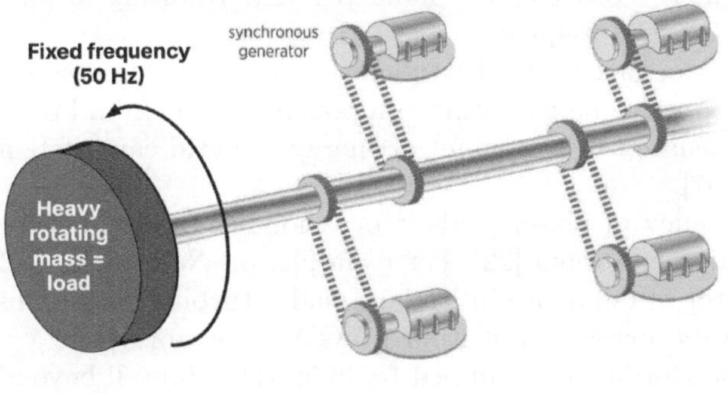

Figure 5.4: Analogy to explain the concept of inertia on electricity grids. *Source: inspired by the report Inertia and the power grid: A guide without the spin, the National Renewable Energy Laboratory* [22].

demand exceeds generation, the frequency of the current will drop, and vice-versa [66]. In the real world, the frequency of a grid is thus never constant; it constantly fluctuates from the random turning on and off of individual loads. Operators only need to ensure that it remains within safe margins, around a nominal value (50 Hz in Europe, 60 Hz in the US), as all electrical components are uniquely designed to function in that range.

Inertia is critical for grid operators in controlling the scale of random fluctuations in frequency. To explain why, let us draw a simplifying analogy. As can be seen in Figure 5.4, a power grid can be simplified as a set of synchronous generators coupled together by a rotating shaft. At the end of that shaft would be placed a heavy mass, rotating in a synchronous fashion. In a stationary regime, the generators would be able to power the rotation of that mass at a set frequency, with their power compensating losses and air resistance. Here, the rotating mass would be analogous to the demand, or load, of the grid [22]. If the load was to decrease, that would equate to instantly reducing the size and mass of the rotating object, which would lead to an increase of the rotation speed, and thus the grid frequency, due to the conservation of angular momentum.

In this system, the larger the rotating mass, the larger the inertia, and thus the smaller the variations of the frequency of the system under a given perturbation. This analogy explains how grid operators can use inertia to minimise the amplitude of eventual frequency perturbations caused by constant changes in consumption and generation [22, 67].

The larger the inertia of a grid, the less noticeable will be the turning on or off of a single light bulb on the frequency of the system. A typical LED light bulb requires around 4 watts of power to operate. Within the Synchronous grid of Continental Europe, where the total load generally exceeds several hundred gigawatts [93], turning on a lamp would only represent an increase of the load in the order of the nano percent, resulting in an imperceptible variation of the grid frequency.

System-wide, variations of frequency can thus be seen as the storage and discharge of kinetic energy within components of the grid. Frequency margins therefore translate to the amount of energy the grid can safely accumulate or discard [22, 57].

The frequency of power grids is occasionally pushed beyond safety margins by unforeseen events [22]. For example, on November 17, 2024, the 3rd nuclear reactor of Olkiluoto in Finland had a turbine failure, instantaneously taking down its power output from 1.6 GW to none [91]. As a result, the frequency of the Nordic grid dropped to 49.59 Hz [52], well beyond the minimal quasi-steady-state frequency of 49.8 Hz [83]. This deviation triggered multiple mechanisms to restore the frequency within operational limits. Grid operators have at their disposal a hierarchy of tools they can use to regulate the frequency, with different reaction times and scale [79, 98]. For example, within seconds after the trip, utility-scale batteries that have stored electricity as chemical energy can start releasing it on the grid [36]; minutes after, pumped hydroelectrical dams can start discharging stored water and interconnections adapt their power flows [53]. As such, ancillary and imbalance markets are used by operators to create a new grid configuration that is able to continue to meet the demand for electricity in the absence of that nuclear reactor.

Following these technical apartés, it is now possible to scrutinize tech strategies to reduce reported emissions.

Corporate carbon accounting rules, as defined by the GHG protocol [86], allow the use of two different accounting methodologies to assess the effect of carbon-aware computing on emissions.

- On the one hand, if a company relies on market-based accounting, it is possible that the implementation of carbon awareness will have no effect on its reported emissions. Today, in most cases, the EACs are not granular enough to account for the time when electricity consumption occurred [12, 54]. If, in fact, certificates are purchased annually to cover all electricity consumption, as is most commonly done today [12, 54], the decision to target preferential consumption times will have no effect on the reported emissions of the company. Only the adoption of granular certificates [54], or the self-imposition of stricter claiming rules [54], will allow carbon awareness to influence the reporting of market-based carbon emissions.

- On the other hand, a company using location-based accounting will see the effect of implementing carbon awareness on its reported emissions. If that company is able to reliably shift its electricity consumption at times of lower carbon intensity, then, by definition, its attributed emissions will be reduced [86].

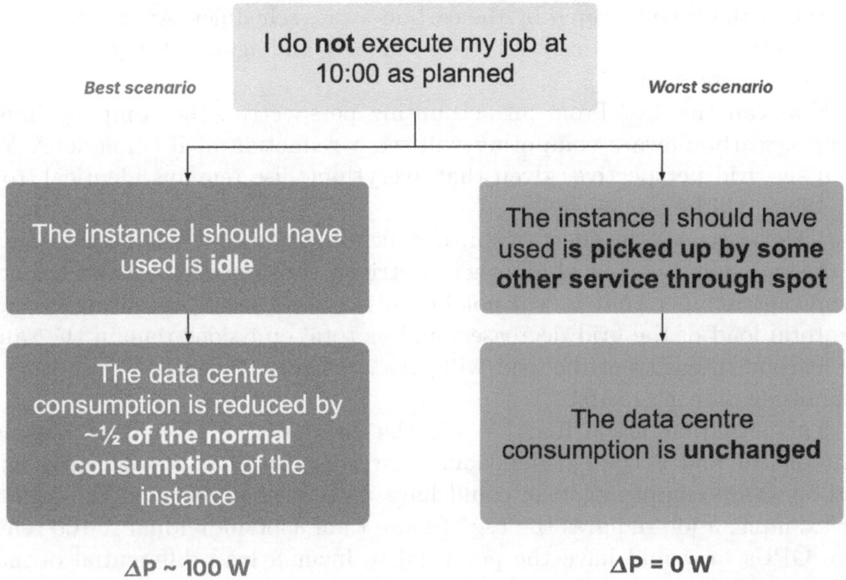

Figure 5.5: Understanding the effect of choosing not to run a computing job on the power consumed from the grid.

Although not inherent to carbon-aware computing, this reveals a potential first problem for individuals trying to implement it. If the company they work for uses exclusively market-based carbon accounting, the effect of the adoption of carbon awareness on real-world emissions reduction will be obscured.

Furthermore, under the apparent simplicity of the answer provided by the location-based accounting lie different layers of complexity that need to be unpacked. Let us take the example of a relatively small company that makes a computing job running in a public cloud carbon-aware. Using some carbon intensity forecast, the schedule of the job will be shifted in time. As depicted in Figure 5.5, at the time when the job should originally have been run, the datacenter will see, at best, a power differential corresponding to that incurred by machines staying idle instead of being utilized [17].

Note that the same would have occurred if the job was to be run on-premise; in that case, it is certain that no other job would have filled in the available capacity.

In the best case, the grid will see a corresponding drop in demand from the datacenter. Given the nominal power outputs of servers, this differential should be within the order of magnitudes of 100 W to 1 kW [21, 71]. Given that the datacenter is located within a large enough synchronous electricity grid, i.e., in Europe, this will result in absolutely no difference for the grid.

Total grid emissions will remain the same during that period. The same will be true at the time targeted by the carbon-aware scheduler. At worst, the grid will experience an unnoticeable increase in load, and again total grid emissions will remain the same.

How can this be? From an accounting perspective, the company implementing carbon-aware computing will see a reduction of its emissions. Yet, from the grid perspective, given that everything else remains identical, total emissions will be unchanged.

This is because the effect of shifting the execution time will be transferred to the emissions account of all other electricity consumers, but in such a non-significant manner that it will not be noticeable. By not executing the job, the total load of the grid decreases, and as total emissions remain the same, the carbon intensity of the grid will ever so slightly increase by an order of magnitude of, e.g., 1e-10.

This situation has a limit. While the exact system boundary is generally unclear and is context-dependent, we can posit that a sufficiently large carbon-aware computing task could have a noticeable effect on the grid. If, for example, a job requires the use of more than 3300 individual A-100 tensor core GPUs, it would have the potential to incur a load differential of more than 1 MW (300 W each) [71]. This scale has probably been reached many times over with the training of large language models. If that differential did not trigger any frequency response (which is not impossible), and therefore a change in total grid emissions, it could lead to a somewhat noticeable change in the carbon intensity of the grid. For a grid operating under a 10 GW load, the addition of 1 MW of load would reduce the carbon intensity of the grid by 0.01%[2], which could start to slightly affect the emissions report of all other consumers.

Therefore, considering that the total power consumption of large supercomputers or hyperscale datacenters can reach dozens, if not hundreds, of megawatts [43], any significant change in their utilization could lead to a load differential in the order of magnitude of 10 MW. That can be enough not only to affect the carbon intensity of the grid, but also to challenge the assumption that total grid emissions will remain the same. Such a load differential can trigger a frequency response from the grid, not on the same scale as if a nuclear reactor was to trip, for example, but still significant enough to trigger moments after a reaction from ancillary services. Note that reports already mention how hyperscaler datacenters disrupt the correct functioning of electricity grids in the US [70]. Although the observed disruptions are very localized and do not affect operators' ability to maintain a stable frequency, they reveal that some

[2]If we denote CI the carbon intensity of the grid before starting the job, we can decompose it as $CI = \frac{E}{10 \cdot 10^9}$ where E are total grid emissions. Considering that the configuration of the grid and its emissions remain identical when the job begins, the new carbon intensity can be determined as $CI' = \frac{E}{10 \cdot 10^9 + 1 \cdot 10^6} = CI \frac{10 \cdot 10^9}{10 \cdot 10^9 + 1 \cdot 10^6}$.

electricity grids are not braced to reliably cover the demand profiles of large datacenters.

This poses a fundamental challenge; making sufficiently large computing tasks carbon-aware can incur complex effects on the grid. The carbon intensity of the electricity at targeted times for consumption might be affected by that very decision; that has an adverse effect on the potential of carbon awareness for emissions reduction.

Marginal emissions are sometimes presented as the solution to this tension [85, 98]. They are supposed to represent the emissions of the electric generator(s) that respond to changes in load on the local grid at a certain time. From this, it is possible to calculate the marginal carbon intensity, which refers to the carbon intensity of the electricity produced by the marginal power plant or plants, which would respond to changes in the grid load [98]. In most grids, fossil fuel-powered plants are more easily dispatched and thus are more often on the margin [69, 98]. With the integration of large amounts of renewable electricity sources, such as solar or wind, curtailment events occur more frequently [89], during which renewable sources can appear on the margin [98]. Curtailment refers here to situations where the amount of power generated by solar or wind exceeds what can be consumed through the grid [68], or more often (80% of the time), when there are congestions on some of the transmission lines.

Accurate forecasts of the marginal carbon intensity would therefore solve the problem we identified earlier, when the scale of carbon-aware operations reaches a point where they can start to influence grid dynamics. Unfortunately, the demonstration of the accuracy of short-term marginal carbon intensity forecasts is highly contested. Grid operators have publicly stated that due to the complex dynamics of electricity grids, accurately determining the marginal power plant, even retrospectively, is not feasible [1, 75]. The premise behind this indeterminism is that marginal emissions are based on non-realized counterfactuals. Stating that at a time t in the past, the marginal carbon intensity of the grid was X gCO_2e/kWh means that it is expected that if a load had been added at that time, the electricity that supplied that extra demand would have come from a source that feeds electricity with a carbon intensity of X gCO_2e/kWh. This did not happen however, and there is thus no reality to match against this expectation. As historical marginal carbon intensity values are not measurable, it follows that, by extension, so are marginal carbon intensity forecasts. A provider of marginal carbon intensity forecasts could claim that they achieve perfect forecasts, without anyone being able to factually disprove that assertion. Existing methodologies are furthermore agnostic of the additional load on the grid [98]. This is potentially a large source of error; a load differential of 10 MW will result in a vastly different reaction from the grid as one of 1 GW. Finally, it should be noted that regulations exclude the usage of emission reporting based on marginal carbon intensities [40, 86]. This would mean that a company that implements carbon-aware computing based on short-term marginal carbon intensity forecasts would use a different carbon signal for optimization and accounting. As these two signals are not necessarily correlated, that could lead that company to actually see its reported emissions increase [30].

This in-depth exploration of the consequences of carbon-aware computing should highlight that its blind implementation, especially using the wrong signal for optimization, should not be considered as a robust mechanism to reduce real-world carbon emissions. The implementation of carbon awareness within an organization should account for the scale of its operations and fit within its existing reporting strategy.

5.4 Blueprint for Sustainable Computing

If carbon awareness alone is not sufficient to enable sustainable computing, what is? It can have an important place in strategies to reduce emissions from computing, when embedded within a broader reflection about sustainability. It is important to understand that shifting the execution of computing jobs is only a step in a hierarchy of initiatives that can result in real-world emission reductions.

Evaluating these initiatives with regards to three criteria – order of magnitude for the amount of emission reduction, certainty of reduction, verifiability of reduction – can help prioritize initiatives within a hierarchy for impact. We thus propose a four-step blueprint for making computing more sustainable, consisting of the following principles: reduce, locate, provide, and optimize.

1. **Reduce** the total amount of energy used by your jobs. Reducing the total amount of energy used to provide the same level of service is the most direct, certain and verifiable way to reduce total emissions. Given that a certain job must run, provisioning the most minimalistic hardware configuration that can safely execute it instead of overprovisioning results in savings in two different ways.

   For the hardware powering servers or virtual machines, most emissions are embodied. They correspond to the emissions resulting from the production and disposal of this equipment [21, 100]. A significant portion of it is resulting from the energy that has been needed in the supply chain to, e.g., smelt metals or manufacture the semiconductors used in the components [14]. This means that using the most efficient hardware to execute a given job will result across the full product lifecycle in less energy being used to provide the same service.

   Furthermore, the amount of power used by hardware does not scale one-to-one with utilization. An idle server will typically use half of its peak power; total power then grows linearly with utilization [41]. Downscaling hardware as much as possible will therefore lead to a decrease in the energy required to run any computing job. Surfacing metrics about utilization is a direct way to verify whether the allocated computing resources are adequate for the tasks at hand. Setting precise targets for utilization, as for example trying to reach 50% of utilization across the board [51], should result in the weeding out of zombies, computing instances that are no longer doing any useful work. Some organizations have found that zombies can represent a significant portion of the machines used [20]. Note that the usage of public cloud services for computing offers easy solutions that implement autoscaling [5]. Due to the cost of hardware, public cloud providers have a direct incentive to increase the utilization of their resources, and the usage of public cloud can therefore be seen as a good way to achieve higher utilization of computing hardware.

Additionally, reflections about the validity of maintaining the same level of service, or computing power at all times can open the door to interesting energy efficiency gains. In accordance with the design philosophy of graceful degradation [18, 50], one could imagine building systems that are able to operate at a sufficient level of service despite some of their parts not being operational because the hardware on which they were running is fully utilized. That would enable us not to provision hardware only to satisfy computing requirements needed in edge cases.

2. **Locate**, consistently, compute in areas where low carbon electricity is abundant and where there are credible plans to make the grid even cleaner.

As exposed earlier, the carbon intensity of the electricity that powers computers varies drastically from region to region. In the US in 2023 for example, the carbon intensity of the grid in the Public Utility District No. 1 of Chelan County (US-NW-CHPD), in the Pacific Northwest region, was on average 23 times lower than within the South Carolina Public Service Authority (US-CAR-SC) [31]. Similar orders of magnitude are also observed in Europe. Electricity in the South Central Sweden region (SE-SE3) was on average 33 times less emissions intensive than in Poland in 2023. These disparities are likely to persist in the early 2030s. The national renewable targets for 2030 of 51% for Poland or 59% for the US [33], and the obstacles that slow the installation of other low-carbon electricity sources [16], show that the gap between grids with low and high carbon intensity will not be bridged in the second part of the 2020s. Some countries or regions of the world are more transparent around the envisioned development pathways for their electrical grids, such as France [78] or Denmark [35]. This creates the opportunity to deliberately run compute-heavy resources in areas where low-carbon electricity is already and should remain available. This is especially relevant for companies and organizations that rely on a location-based accounting of their emissions. It also affects proponents of market-based accounts, as the procurement of RECs in low-carbon intensity grids will become comparatively easier as their granularity increases and as regulations start to embed deliverability criteria [96].

Cloud providers are already publicly disclosing information about the carbon content of the grids that power their datacenters [48]. Encouraging the displacement of computing to such regions of the world could negatively affect the stability of the grids located there. As exposed earlier, the large-scale installation of datacenters on some grids can either lead to grid instabilities [70] or to decarbonization impediments [88]. Targeting areas abundant in low-carbon electricity within large capacity grids for computing services can mitigate this risk. Grids of Québec, or France, could constitute such examples [31].

3. **Provide** computing power from companies that power their operations with additional low carbon electricity.

Choosing the right provider for computing services can also make a decisive difference for organizations interested in making their activities more sustainable. Different providers will offer a portfolio of services whose carbon intensity differs substantially. For example, moving computing jobs from a provider operating exclusively in North Virginia to one operating in Iceland would have a significant influence on an organization's location-based carbon footprint [31]. Providers, however, do not only differ in terms of where they operate their datacenters. The energy procurement strategies of computing services providers shine a light on their commitment to decarbonizing their operations. Some providers will set targets to purchase only electricity from renewable sources to power their operations [4, 61, 74]. Among these, the criteria for the procurement of clean electricity vary. Some are non-prescriptive about the renewable energy certificates that can be used [4]; others put emphasis on ensuring additionality for the renewable energy purchased [61]; others also include criteria for the temporal and geographical matching of renewable claims with generation [19, 97]. Transparency around these criteria is key, as they ultimately determine how credible these strategies are [12], transparent providers generally being the most committed to decarbonizing their energy supply [72]. An organization desiring to minimize its environmental impact can therefore analyze the energy procurement strategies of potential providers and include their alignment with its own decarbonization pathway in the selection process. Some examples exist of corporations changing cloud providers on that basis [11]. In the future, other criteria could be used to assess whether datacenter operators support grid decarbonization efforts. Early studies [2, 44] suggest that datacenters could, for example, be used to provide ancillary services to help with frequency regulation, and some implementations are already emerging [34].

Another point to consider when choosing a provider of computing services is whether they offer actionable tools to reduce one's footprint. Some public cloud providers offer customers carbon footprint reports that include both location-based and market-based emissions [47], while others only provide market-based accounts [6]. The latter can be of limited practical use when providers claim that their market-based emissions near 0. The emissions report must be granular and not only top-down to be actionable. Without these criteria, it would be, for example, impossible for users to verify that moving their computations to a cleaner grid, or adopting an auto-scaling mechanism, would reduce their footprint.

4. **Optimize** your compute to match times of available low-carbon electricity.

Finally, optimizing the execution time of jobs is the last lever to reduce the footprint of computing. Some tools already exist that leverage external forecasts to shift the execution of software to match the availability of low carbon power [84]. Using the right forecast for optimization is of critical importance, as different signals might suggest shifting executions at significantly different times. While we have detailed earlier why using short-term marginal carbon intensity forecasts is not advisable, several options are left for optimization. Although targeting times of minimal grid net load (obtained by removing wind and solar generation from the total load) can be helpful in avoiding overloading grids, this might not lead to the greatest reduction in location-based emissions. In fact, there are regions of the world, such as Sweden and Norway, where the flow traced carbon intensity of the grid is lowest at times of high load as hydropower is dispatched to meet the additional demand [31]. Targeting the share of renewable electricity, or the total wind and solar generation, is also an intuitive option that creates a long-term incentive to install low carbon capacity but might not be the most meaningful metric in regions with low penetration of renewables. Finally, using forecasts of the flow traced carbon intensity is the most direct method to ensure a location-based reduction in one's emissions from computing through load shifting.

5.5 References

[1] 50 Hertz. eCO_2 grid How we calculate. `https://eco2grid.50hertz.com/calculation`, 2024. [Online; accessed 25-January-2025].

[2] D. Aikema, R. Simmonds, and H. Zareipour. Delivering ancillary services with data centres. *Sustainable Computing: Informatics and Systems*, 3(3):172–182, 2013. Selected papers from the 2012 IEEE International Green Computing Conference (IGCC 2012).

[3] Amazon. Amazon sustainability report. Technical report, Amazon, 2023.

[4] Amazon. Renewable energy methodology. Technical report, Amazon, 2024.

[5] Amazon Web Services. AWS Auto Scaling. `https://aws.amazon.com/autoscaling/`, 2025. [Online; accessed 25-January-2025].

[6] Amazon Web Services. Customer carbon footprint tool. `https://aws.amazon.com/aws-cost-management/aws-customer-carbon-footprint-tool/`, 2025. [Online; accessed 25-January-2025].

[7] Association of Issuing Bodies (AIB). Renewable energy guarantees of origin. `https://www.aib-net.org/certification/certificates-supported/renewable-energy-guarantees-origin`, 2025. [Online; accessed 25-January-2025].

[8] S. Ayers, S. Ballan, V. Gray, and R. McDonald. Measuring the emissions and energy footprint of the ict sector : Implications for climate action. Technical report, World Bank, 2024.

[9] W. Bank. *World development report 2021*. World Development Report. World Bank Publications, Washington, D.C., DC, July 2021.

[10] W. Bank. *Green digital transformation: How to sustainably close the digital divide and harness digital tools for climate action*. Washington, DC: World Bank, Nov. 2023.

[11] C. Bohic. Migration Cloud: comment Back Market est passé d'AWS à Google Cloud. *Silicon*, 06 2024. `https://www.silicon.fr/Thematique/cloud-1370/Breves/Migration-Cloud-comment-Back-Market-est-passe-d-AWS-a-460615.htm`, [Online; accessed 01-December-2025.

[12] K. Bryan, C. Hodgson, and J. Tauschinski. Big tech's bid to rewrite the rules on net zero. *The Financial Times*, 2024. `https://www.ft.com/content/2d6fc319-2165-42fb-8de1-0edf1d765be3`, [Online; accessed 01-December-2025].

[13] W. Buchanan, J. Foxon, D. Cooke, S. Iyer, E. Graham, B. DeRusha, C. Binder, K. Chiu, L. Corso, H. Richardson, V. Knight, A. Hussain, A. Allison, and N. Mathews. Carbon-aware computing measuring and reducing the carbon intensity associated with software in execution. Technical report, The Green Software Foundation, 2023.

https://news.microsoft.com/wp-content/uploads/prod/sites/418/2023/01/carbon_aware_computing_whitepaper.pdf.

[14] A. Busa and M. Hegeman. Life Cycle Assessment of Dell R740+. Technical report, Thinkstep, 2019. https://www.delltechnologies.com/asset/en-us/products/servers/technical-support/Full_LCA_Dell_R740.pdf.

[15] J. Canadell, P. Monteiro, M. Costa, L. Cotrim da Cunha, P. Cox, A. Eliseev, S. Henson, M. Ishii, S. Jaccard, C. Koven, A. Lohila, P. Patra, S. Piao, J. Rogelj, S. Syampungani, S. Zaehle, and K. Zickfeld. Global carbon and other biogeochemical cycles and feedbacks. In V. Masson-Delmotte, P. Zhai, A. Pirani, S. L. Connors, C. Péan, S. Berger, N. Caud, Y. Chen, L. Goldfarb, M. I. Gomis, M. Huang, K. Leitzell, E. Lonnoy, J. B. R. Matthews, T. K. Maycock, T. Waterfield, O. Yelekçi, R. Yu, and B. Zhou, editors, *Climate Change 2021: The Physical Science Basis. Contribution of Working Group I to the Sixth Assessment Report of the Intergovernmental Panel on Climate Change*, book section 5, pages 673–815. Cambridge University Press, Cambridge, UK and New York, NY, USA, 2021.

[16] P. Carajilescov and J. M. L. Moreira. Construction time of PWRs. In *International Nuclear Atlantic Conference*. INAC, 2011.

[17] A. Cockcroft. Don't follow the sun: Scheduling compute workloads to chase green energy can be counter-productive, 2023. https://adrianco.medium.com/dont-follow-the-sun-scheduling-compute-workloads-to-chase-green-energy-can-be-counter-productive-b0cde6681763, [Online; accessed 01-December-2025].

[18] Code Reliant. Failing with dignity: A deep dive into graceful degradation. https://www.codereliant.io/failing-with-dignity/-us/azure/azure-monitor/autoscale/autoscale-overview, 2023. [Online; accessed 25-January-2025].

[19] H. Crammer and S. Goodman. Accelerating a carbon-free future with hourly energy tracking. *Google Blog*, 2023. https://cloud.google.com/blog/topics/sustainability/t-eacs-show-promise-for-helping-decarbonize-the-grid.

[20] A. Currie, S. Hsu, and S. Bergman. *Building green software*. O'Reilly Media, Sebastopol, CA, Mar. 2024.

[21] B. Davy. Building an AWS EC2 Carbon Emissions Dataset, 2021. https://medium.com/teads-engineering/building-an-aws-ec2-carbon-emissions-dataset-3f0fd76c98ac#e38a.

[22] P. Denholm, T. Mai, R. W. Kenyon, B. Kroposki, and M. O'Malley. Inertia and the power grid: A guide without the spin. Technical report, National Renewable Energy Laboratory (NREL), 2020.

[23] EDF. Want to know more about choosing your electricity source?, 2024. https://www.edfenergy.com/large-business/want-know-more-about-choosing-your-electricity-source, [Online; accessed 25-January-2025].

[24] Electricity Maps. How to trace back the origin of electricity. https://www.electricitymaps.com/blog/flow-tracing, 2021. [Online; accessed 25-January-2025].

[25] Electricity Maps. Understanding electricity scope 2 attribution rules. https://www.electricitymaps.com/reports-and-guides/accounting-guide, 2022. [Online; accessed 25-January-2025].

[26] Electricity Maps. Consumption vs production: Why grid emissions should be calculated with flow-traced data. https://www.electricitymaps.com/blog/consumption-carbon-intensity, 2023. [Online; accessed 25-January-2025].

[27] Electricity Maps. What is additionality (and emissionality)? https://www.electricitymaps.com/blog/what-is-additionality, 2023. [Online; accessed 25-January-2025].

[28] Electricity Maps. Carbon aware computing. https://www.electricitymaps.com/industries/carbon-aware-computing, 2024. [Online; accessed 25-January-2025].

[29] Electricity Maps. Electricity maps methodology. https://www.electricitymaps.com/methodology#carbon-intensity-and-emission-factors, 2024. [Online; accessed 25-January-2025].

[30] Electricity Maps. Marginal signals explained (Part 6): Optimizing electricity consumption with a marginal signal may not reduce its carbon footprint, 2024. https://www.electricitymaps.com/blog/optimizing-electricity-consumption-with-a-marginal-signal-may-not-reduce-its-carbon-footprint, [Online; accessed 25-January-2025].

[31] Electricity Maps. Electricity maps - global hourly electricity mixes. https://app.electricitymaps.com/map/72h/hourly, 2025. [Online; accessed 25-January-2025].

[32] Electricity Maps. Electricity maps - yearly average electricity mix for the repulic of ireland. https://app.electricitymaps.com/zone/IE/all/yearly/2024-01-01T00:00:00.000Z#origin_chart, 2025. [Online; accessed 25-January-2025].

[33] Ember. 2030 global renewable target tracker. https://ember-energy.org/data/2030-global-renewable-target-tracker/, 2025. [Online; accessed 25-January-2025].

[34] Enel x. How data centers support the power grid with ancillary services? https://www.enelx.com/tw/en/resources/how-data-centers-support-grids, 2025. [Online; accessed 25-January-2025].

[35] Energinet. Long-term development needs in the power grid energinet's long-term development plan 2022 – needs analysis. Technical report, Energinet, 2022.

[36] Energinet. Outlook for ancillary services 2023-2040. Technical report, Energinet, 2023.

[37] Energy-charts. Homepage. https://www.energy-charts.info/charts/power/chart.htm?c=DE, 2025. [Online; accessed 25-January-2025].

[38] Entelios. Anciliary services. https://www.entelios.com/en/ancillaryservices, 2025. [Online; accessed 25-January-2025].

[39] Entergy. Meta selects Entergy, Northeast Louisiana as site of $10B data center. https://www.entergynewsroom.com/news/meta-selects-northeast-louisiana-as-site-10-billion-data-center/, 2025. [Online; accessed 25-January-2025].

[40] European Union (EU). Directive (EU) 2018/2001 of the European Parliament and of the Council by establishing a Union methodology setting out detailed rules for the production of renewable liquid and gaseous transport fuels of non-biological origin. *Official Journal of the European Union*, L 157/11, 2023.

[41] X. Fan, W.-D. Weber, and L. Barroso. Power provisioning for a warehouse-sized computer. *ACM Sigarch Computer Architecture News*, 35:13–23, 06 2007.

[42] T. G. W. Foundation. Grid aware websites. https://www.thegreenwebfoundation.org/tools/grid-aware-websites/, 2024. [Online; accessed 25-January-2025].

[43] I. Fried. Up close with the world's largest supercomputer. *Axios*, 2025. https://www.axios.com/2025/01/10/lawrence-livermore-lab-supercomputer, [Online; accessed 01-December-2025].

[44] M. Ghamkhari and H. Mohsenian-Rad. Data centers to offer ancillary services. In *2012 IEEE Third International Conference on Smart Grid Communications (SmartGridComm)*, pages 436–441, 2012.

[45] Google. Google 2024 environmental report. Technical report, Google, 2024. https://www.gstatic.com/gumdrop/sustainability/google-2024-environmental-report.pdf.

[46] Google. Data centers, efficiency. https://www.google.com/about/datacenters/efficiency/, 2025. [Online; accessed 25-January-2025].

[47] Google Cloud. Carbon footprint. https://cloud.google.com/carbon-footprint, 2025. [Online; accessed 25-January-2025].

[48] Google Cloud. Carbon free energy for google cloud regions. https://cloud.google.com/sustainability/region-carbon, 2025. [Online; accessed 25-January-2025].

[49] GridStatus. Homepage. https://www.gridstatus.io/, 2025. [Online; accessed 25-January-2025].

[50] C. Humble. What are demand shifting and shaping in green software? https://www.conissaunce.com/demand-shifting-and-shaping.html, 2024. [Online; accessed 25-January-2025].

[51] C. Humble. Writing greener software even when you are stuck on-prem. https://www.conissaunce.com/presentations.html#writing-greener-software, 2024. [Online; accessed 25-January-2025].

[52] Hybrid Greentech. Nordic Nuclear Power Plant Disturbances: Olkiluoto and Forsmark Frequency Events, November 2024. LinkedIn Post https://www.linkedin.com/posts/hybridgreentech_the-finish-nuclear-power-plant-olkiluoto-activity- 7264352376675835905-TDCa. Discusses multiple frequency drop incidents at Olkiluoto and Forsmark nuclear power plants in 2023-2024.

[53] S. Hänninen, C. Evens, P. Divshali, J. Millar, M. Lehtonen, and T. Salnikova. Survey on power system ancillary services. Technical report, Energiforsk, 2020.

[54] T. E. T. Initiative. Energytag and granular energy certificates: Accelerating the transition to 24/7 clean power. Technical report, The Energy Tag Initiative, 2021.

[55] International Renewable Energy Agency (IRENA). Additionality principle. https://www.irena.org/Innovation-landscape-for-smart-electrification/Power-to-hydrogen/10-Additionality-principle, 2023. [Online; accessed 25-January-2025].

[56] F. Irani. Carbon aware site. https://fershad.com/carbon-aware-site/, 2023. [Online; accessed 25-January-2025].

[57] B. J. Kirby. Frequency regulation basics and trends. Technical report, Oak Ridge National Laboratory, 2004.

[58] R. Koningstein. We now do more computing where there's cleaner energy. https://blog.google/outreach-initiatives/sustainability/carbon-aware-computing-location/, 2021. [Online; accessed 25-January-2025].

[59] H. Ma, Z. Zhou, X. Zhang, and X. Chen. Toward carbon-neutral edge computing: Greening edge ai by harnessing spot and future carbon markets. *IEEE Internet of Things Journal*, 10(18):16637–16649, 2023.

[60] J. Mellerud. Bitcoin mining as a demand response in an electric power system: A case study of the ERCOT-system in Texas. Master's thesis, Nord University, 2021.

[61] Meta. Meta sustainability report - for a better reality. Technical report, Meta, 2024. https://sustainability.atmeta.com/wp-content/uploads/2024/08/Meta-2024-Sustainability-Report.pdf.

[62] Meta. Accelerating the Next Wave of Nuclear to Power AI Innovation. https://sustainability.atmeta.com/blog/2024/12/03/accelerating-the-next-wave-of-nuclear-to-power-ai-innovation/, 2025. [Online; accessed 25-January-2025].

[63] H. Momblanco, K. Aiuto, and S. Huckins. Overview of GHG Protocol Integration in Regulatory Climate Disclosure Rules. Technical report, Greenhouse Gas Protocol, 2024.

[64] National Energy System Operator (NESO). How do we balance the grid? https://www.neso.energy/energy-101/electricity-explained/how-do-we-balance-grid, 2025. [Online; accessed 25-January-2025].

[65] National Energy System Operator (NESO). NESO carbon intensity API. https://carbonintensity.org.uk/, 2025. [Online; accessed 25-January-2025].

[66] National Energy System Operator (NESO). What is frequency? https://www.neso.energy/energy-101/electricity-explained/how-do-we-balance-grid/what-frequency, 2025. [Online; accessed 25-January-2025].

[67] National Energy System Operator (NESO). What is inertia? https://www.neso.energy/energy-101/electricity-explained/how-do-we-balance-grid/what-inertia, 2025. [Online; accessed 25-January-2025].

[68] Next Kraftwerke. What is curtailment of electricity? https://www.next-kraftwerke.com/knowledge/curtailment-electricity, 2024. [Online; accessed 25-January-2025].

[69] Next Kraftwerke. What is the merit order curve in the power system? https://www.next-kraftwerke.be/knowledge-hub/merit-order-curve, 2024. [Online; accessed 25-January-2025].

[70] L. Nicoletti, N. Malik, and A. Tartar. Ai needs so much power, it's making yours worse. *Bloomberg*, 2024. https://www.bloomberg.com/graphics/2024-ai-power-home-appliances/.

[71] Nvidia. NVIDIA A100 Tensor Core GPU Datasheet. Technical report, NVIDIA, 2022. https://www.nvidia.com/content/dam/en-zz/Solutions/Data-Center/a100/pdf/nvidia-a100-datasheet-nvidia-us-2188504-web.pdf.

[72] D. Oberhaus. Amazon, Google, Microsoft: Here's Who Has the Greenest Cloud. *Wired*, 2019. https://www.wired.com/story/amazon-google-microsoft-green-clouds-and-hyperscale-data-centers/.

[73] Persefoni. Scope 2 accounting: Market-based vs. location-based methods explained. https://www.persefoni.com/blog/scope-2-dual-reporting-market-and-location-based-carbon-accounting, 2025. [Online; accessed 25-January-2025].

[74] S. Pichai. Our third decade of climate action: Realizing a carbon-free future. *Google Blog*, 2020. https://blog.google/outreach-initiatives/sustainability/our-third-decade-climate-action-realizing-carbon-free-future/. [Online; accessed 01-December-2025].

[75] PJM. Marginal emissions rate – a primer. Technical report, PJM, 2025. https://www.pjm.com/-/media/DotCom/etools/data-miner-2/marginal-emissions-primer.ashx.

[76] Reel. What is additionality? https://www.reel.energy/post/what-is-additionality, 2024. [Online; accessed 25-January-2025].

[77] H. Ritchie. What share of global CO_2 emissions come from aviation? *Our World in Data*, 2024. https://ourworldindata.org/global-aviation-emissions.

[78] Réseau de Transport d'Électricité (RTE). Energy pathways to 2050 key results. Technical report, Réseau de Transport d'Électricité (RTE), 2021. https://assets.rte-france.com/analyse-et-donnees/2023-07/Energy%20pathways%202050_Key%20results.pdf.

[79] Réseau de Transport d'Électricité (RTE). Providing frequency ancillary services. https://www.services-rte.com/en/learn-more-about-our-services/providing-frequency-ancillary-services.html, 2024. [Online; accessed 25-January-2025].

[80] Singularity. Homepage. https://singularity.energy/, 2025. [Online; accessed 25-January-2025].

[81] Singularity. Homepage. https://watttime.org/, 2025. [Online; accessed 25-January-2025].

[82] M. Sotos. GHG Protocol Scope 2 Guidance Executive Summary. Technical report, The Greenhouse Gas Protocol, 2022. https://ghgprotocol.org/sites/default/files/2022-12/Scope2_ExecSum_Final.pdf. [Online; accessed 01-December-2025].

[83] The European Network of Transmission System Operators for Electricity (ENTSO-E). Policy 1: Load-frequency control and performance. Technical report, ENTSO-E, 2009.

[84] The Green Software Foundation. Carbon Aware SDK. https://carbon-awaresdk.greensoftware.foundation/, 2023. [Online; accessed 25-January-2025].

[85] The Green Software Foundation. Carbon awareness. https://learn.greensoftware.foundation/carbon-awareness/, 2023. [Online; accessed 25-January-2025].

[86] The Greenhouse Gas Protocol. A corporate accounting and reporting standard revised edition. Technical report, The Greenhouse Gas Protocol, 2004.

[87] The Greenhouse Gas Protocol. Corporate value chain (scope 3) accounting and reporting standard. Technical report, The Greenhouse Gas Protocol, 2011.

[88] The Internation Energy Agency (IEA). Electricity 2024. analysis and forecast to 2026. https://www.iea.org/reports/electricity-2024, 2024. [Online; accessed 25-January-2025].

[89] The International Energy Agency. Will more wind and solar PV capacity lead to more generation curtailment?, 2023. https://www.iea.org/reports/renewable-energy-market-update-june-2023/will-more-wind-and-solar-pv-capacity-lead-to-more-generation-curtailment, [Online; accessed 25-January-2025].

[90] B. Tranberg, O. Corradi, B. Lajoie, T. Gibon, I. Staffell, and G. B. Andresen. Real-time carbon accounting method for the european electricity markets. *Energy Strategy Reviews*, 26:100367, 2019.

[91] TVO. Electricity production at Olkiluoto 3 was interrupted. https://www.tvo.fi/en/index/news/pressreleasesstockexchangereleases/2024/electricityproductionatolkiluoto3wasinterrupted.html, 2024. [Online; accessed 25-January-2025].

[92] UCTE. Final Report System Disturbance on 4 November 2006. Technical report, Union for the Coordination of Transmission of Electricity (UCTE), 2006. https://ecolo.org/documents/documents_in_e

nglish/blackout-nov-06-UCTE-report.pdf. [Online; accessed 01-December-2025].

[93] UCTE. Annual report 2008 on the move. Technical report, Union for the Coordination of Transmission of Electricity (UCTE), 2008. https://eepublicdownloads.entsoe.eu/clean-documents/pre2015/publications/ce/report_2008.pdf. [Online; accessed 01-December-2025].

[94] United States Environmental Protection Agency (EPA). Energy Attribute Certificates (EACs). https://www.epa.gov/green-power-markets/energy-attribute-certificates-eacs, 2025. [Online; accessed 25-January-2025].

[95] Unknown. Pearl street central station. *Thomas A. Edison Papers*, 1882. https://edison.rutgers.edu/images/innovations/EL/PearlStreetHeadnote.pdf. [Online; accessed 01-December-2025].

[96] U.S. Department of the Treasury. U.S. Department of the Treasury, IRS Release Guidance on Hydrogen Production Credit to Drive American Innovation and Strengthen Energy Security. https://home.treasury.gov/news/press-releases/jy2010, 12 2023.

[97] N. Walsh. Achieving 100 percent renewable energy with 24/7 monitoring in Microsoft Sweden. *Azure Blog*, 2020. https://azure.microsoft.com/en-us/blog/achieving-100-percent-renewable-energy-with-247- monitoring-in-microsoft-sweden/. [Online; accessed 01-December-2025].

[98] WattTime. SIGNAL: Marginal CO_2. https://watttime.org/data-science/data-signals/marginal-co2/, 2024. [Online; accessed 25-January-2025].

[99] J. Weatherbed. This Windows 11 update is trying to save the world, 2022. https://www.techradar.com/news/windows-11-is-getting-an-eco-friendly-update-but- could-microsoft-do-more, [Online; accessed 25-January-2025].

[100] G. Wengel. A guide to carbon-aware computing. https://www.climatiq.io/blog/a-guide-to-carbon-aware-computing, 2023. [Online; accessed 25-January-2025].

[101] Xbox. Xbox is now the first carbon aware console, update rolling out to everyone soon. https://news.xbox.com/en-us/2023/01/11/xbox-carbon-aware-console-sustainability/, 2023. [Online; accessed 25-January-2025].

Part III

Power Optimization for Job Execution

Chapter 6

Leveraging Application Characteristics for Energy Savings

Amina Guermouche and Piotr Luszczek

This chapter provides insights and studies into power draw and energy consumption of a variety of scientific applications. Multiple aspects are considered, including power monitoring, capping, and predicting. This allowed us to achieve improved scheduling in the studied use cases. From the standpoint of reducing the energy consumption, it was possible to regulate frequency and/or power in order to minimize the effects of performance degradation. This involved combining the essential application characteristics and fully leveraging them for controlled influence on performance/energy trade-offs. Some of the observed effects were counterintuitive due to complex interactions between modern scientific applications and the advanced hardware platforms.

Acknowledgements

Research was sponsored by the Department of the Air Force Artificial Intelligence Accelerator and was accomplished under Cooperative Agreement Number FA8750-19-2-1000. The views and conclusions contained in this chapter are those of the authors and should not be interpreted as representing the official policies, either expressed or implied, of the Department of the Air Force or the US government. The US government is authorized to reproduce and distribute reprints for government purposes notwithstanding any copyright notation herein.

6.1 Energy and Power: Monitoring, Regulating, and Scheduling

Traditionally, performance monitoring was conducted at runtime by the Performance Monitoring Unit (PMU) that was introduced during the transition from 32-bit to 64-bit of the majority of consumer- and server-grade Central Processing Units (CPUs). This was introduced to the mass market with IA-64 and AMD64 hardware platforms but eventually the PMU was included in the majority of 32-bit processors, thus allowing the users to track microarchitectural events such as pipeline stalls, system bus transfers, cache misses, and branch mispredictions. However, it is worth noting that these advanced features were introduced gradually and were not present in the early designs.

Later on, the functional scope of the monitoring hardware has expanded, not just to cover more events, but also to account for energy consumption through modeling and even power-limiting features allowing capping of the power draw of various CPU components. This happened with the introduction of the Running Average Power Limit (RAPL) design in Intel Sandy Bridge processors [9], that turned the performance monitoring from read-only to read-and-write system allowing the software layer to take a more active role in influencing the microarchitectural inner-workings of the processors. It was much more versatile than the prior vendor-provided controls over hardware such as Dynamic Voltage and Frequency Scaling (DVFS) [22, 24, 31].

With the growing importance of hardware accelerators, mainly Graphics Processing Units (GPUs), performance monitoring and, later, power capping became a mainstay on all computational devices inside server and desktop systems. NVIDIA Management Library (NVML) is an example of a software middleware for both monitoring and constraining the hardware function at runtime. This enabled a large body of work, including AI-oriented studies in energy/power optimization at scale [11, 23, 25, 29, 41], auto-tuning approaches using machine learning across scientific simulations' kernels [28], and application-specific energy efficiency efforts [33].

6.2 Motivating Example

We begin with a simple case of improving performance by extending a runtime task scheduler with an ability to regulate power and/or frequency of some of the CPU cores. Figure 6.1 shows the potential of such an improvement on the first release of the widely available hardware offering power capping: Intel Sandy Bridge. The difference between the high- and low-power curves indicates about 20% increase in time-to-solution. This inspires an idea to bound the

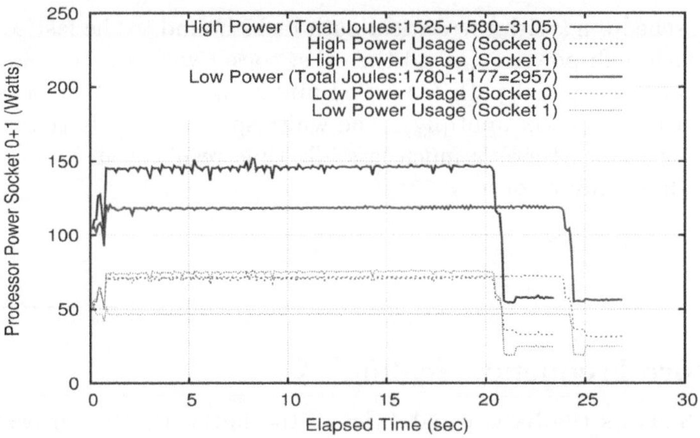

Figure 6.1: Two opposing power states on Intel Sandy Bridge capping core power with frequency by writing to MSR (Model-Specific Register). High-power cap reaching about 150 W and 75 W (solid and dotted lines) for cores reserved for high-priority tasks that finish early at slightly over 20 seconds, and low power of about 120 W and 70 W (solid and dotted lines) for tasks outside of the critical path that finish later at nearly 25 seconds.

critical path tasks to the high-power cores and preferentially schedule the remaining tasks on the low-power cores.

Figure 6.2 shows two runtime traces of tasks when the scheduling is transitioned from identical power settings in the top trace and varied power settings in the bottom, on two-socket Intel Sandy Bridge system with four cores in each processor. The top trace shows the white space around the critical path tasks (panel factorizations of the Cholesky algorithm). The bottom trace shows the

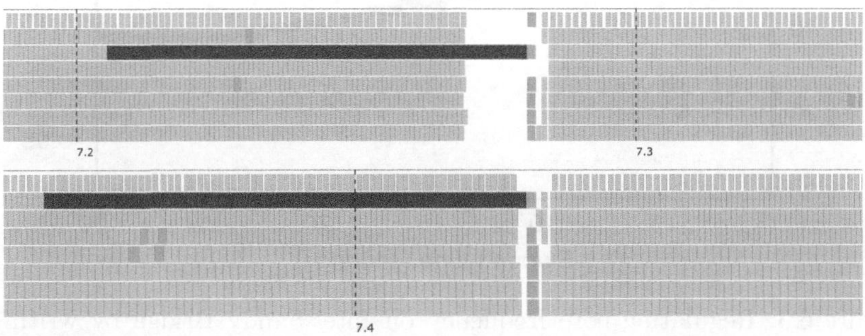

Figure 6.2: Two task schedules: equal power for both sets of cores (top), and devoting full power cores for critical tasks only (bottom). Tasks from the critical path run exclusively on cores without frequency cap.

improved scheduling with the critical path tasks bound to the fast set of cores running with high-power cap, and the other socket with a lower cap handling the tasks off the critical path (so-called update kernels that are mostly specialized forms of matrix multiply). The white space is greatly reduced in the second trace. Thus, there is much less idle time resulting in a more efficient use of the hardware resources.

6.3 Core Frequency Scaling

Figure 6.3 shows the basic capabilities of the initial RAPL hardware in the long discontinued Intel Sandy Bridge CPU [9]. The first major issue of note is a relatively low sampling rate at about 2 Hz. While the internal operation of RAPL is proprietary information, it is known that adjusting the core frequency and voltage regulators takes substantially longer than a few clock cycles. This limits, even in today's CPUs, the speed at which the user is able to change the power cap settings and subsequently read out the energy consumption. Another aspect clearly visible in the figure is the non-linear relation between the

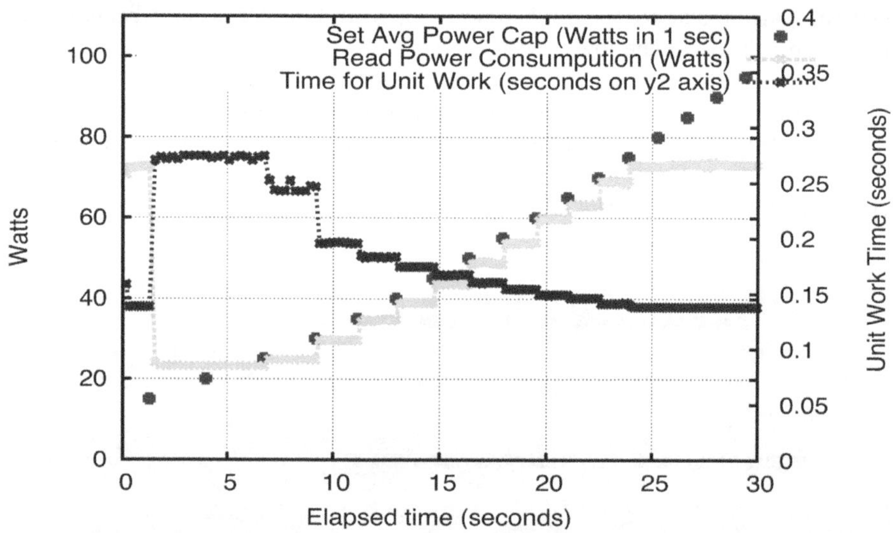

Figure 6.3: Regulating core frequency on Intel Sandy Bridge by writing to MSR (Model-Specific Register) to influence power cap in watts (dark dots) and corresponding readout of power consumption (light crosses) and the resulting reduction in delay of performing computational work or work-delay product (dark crosses).

power cap and the achieved work-delay product [15]. These specific features saw only a limited improvement since the early Intel Sandy Bridge hardware, as we show in the following study.

6.4 Uncore Frequency Scaling

Uncore [19] frequency is the frequency of the Level 3 cache and the memory controllers. In older x86-64 architectures, like Intel Sandy and Ivy Bridge, it was considered as the same domain as core frequency. Intel Haswell architectures introduced different domains for core and uncore frequency, and they can be managed separately since then [17, 20].

Figure 6.4 shows the impact of varying the uncore frequency on the execution time and power consumption of three different applications (described in Section 6.4.2.2). The applications have different behaviors regarding their computational intensity: while High-Performance LINPACK (HPL) and Embarrassingly Parallel (EP) are compute-intensive, Conjugate Gradient (CG) is memory-intensive. Moreover, EP has almost no memory operations.

From Figure 6.4, one can see that varying the uncore frequency has an impact on both HPL and CG. As a matter of fact, as the uncore frequency increases, the performance of both applications improves. Furthermore, for HPL, we even observe a performance improvement of 1.47% compared to the default behavior when reducing the uncore frequency. This is because HPL has a high power consumption that reaches thermal design power (TDP). When we manually decrease the uncore frequency, the power consumption decreases and the core frequency is automatically increased. This behavior will be further discussed in Section 6.4.4. Regarding the power consumption of CG and HPL, both applications see an increase in their power draw as the uncore frequency is decreased. A plateau is observed for HPL for the

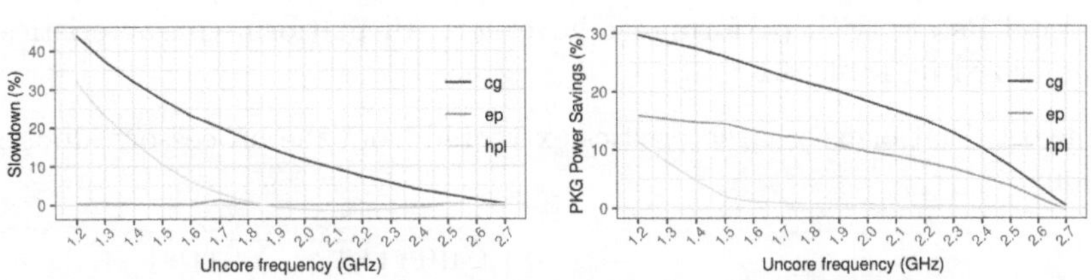

(a) Slowdown (b) PKG power

Figure 6.4: Uncore frequency impact on execution time and package power consumption, on a machine equipped with two Intel Xeon E5-2680 v4 CPUs.

same reasons as described above: because of TDP, we cannot set the uncore frequency when running HPL to more than 2.4 GHz.

Regarding EP, one can see that the uncore frequency has no impact on its performance while providing power savings of 16%.

As a conclusion, Figure 6.4 indicates that while uncore frequency can provide power savings with no or limited impact on performance, it is not well managed on some architectures. For this reason, we proposed a new tool, able to adapt the uncore frequency to the application needs.

6.4.1 Dynamic uncore frequency algorithm

The Dynamic Uncore Frequency (DUF) [1] is an algorithm able to: (i) adapt the uncore frequency to the application needs, and (ii) accept and respect a user-defined tolerated slowdown. For that purpose, DUF periodically reads information about the application being executed, namely its performance and memory bandwidth. At each period, it decides whether it would be beneficial to decrease the uncore frequency based on the decision of the previous period, or if the previous decision negatively impacted the performance. If the performance is not impacted, DUF also checks whether the memory bandwidth did not drop. The goal is to always respect the user-defined tolerated slowdown. The uncore frequency may also be reset if the behavior of the application changes (from compute- to memory-bound or the opposite, or a high difference in flop/s).

Note that DUF operates at the socket level: one instance of DUF runs per socket, and the decisions are made independently based on the measurements on each socket.

6.4.2 Experimental setup

This section presents the target architectures and applications.

6.4.2.1 Hardware settings

We used two servers from the Grid'5000 [4] platform. All platforms run under Intel Pstate with performance governor. All platform characteristics are summarized in Table 6.1.

Table 6.1: Platforms' characteristics extracted from the processors' documentation.

	Chifflet	Yeti
Number of cores	28	64
Microarchitecture	Broadwell	Skylake
TDP (W) per socket	120	125
Uncore frequency (GHz)	[1.2-2.7]	[1.2-2.4]

- CHIFFLET is equipped with two Intel Xeon E5-2680 v4 CPUs (Broadwell microarchitecture) with 14 cores per CPU, and 768 GiB of memory. The uncore frequency ranges from 1.2 GHz to 2.7 GHz. We used chifflet-1 for our experiments.

- YETI is equipped with four Intel Xeon Gold 6130 CPUs (Skylake microarchitecture) with 16 cores per CPU. Each NUMA node has 64 GiB of memory. The uncore frequency ranges from 1.2 GHz to 2.4 GHz. In all experiments, we used yeti-2.

6.4.2.2 Target applications

- The NAS Parallel Benchmarks [2] provide a set of small applications. We use: BT, CG, EP, FT, LU, MG, SP, UA from NPB-3.3.1 OpenMP version. We choose the problem size so that each application execution time is in the [20s-400s] range. On CHIFFLET, EP, MG, and FT run using the class D problem size. The remaining benchmarks run using class C. On YETI, all benchmarks run using class D except SP for which we use class C. The OpenMP threads are bound to cores in a round-robin fashion.

- High-Performance LINPACK (HPL) [10, 26] is a software package that solves a dense linear algebra system of equations. We use HPL version 2.3 compiled with Math Kernel Library (MKL) version 2019.1.144. HPL uses a configuration file where we set NB to 224 on all platforms. N is set to 62720 on CHIFFLET and 91840 on YETI. MPI process grid (P-by-Q) is set to 4-by-7 on CHIFFLET and 8-by-8 on YETI.

- LAMMPS[1] [27] performs molecular dynamics simulation. We use input file `in.lj` provided for the accelerate suite where we set the `run` value to 100,000.

- NwChem[2] [37] is a computational chemistry application. We use the input data set `3carbo.nw` from the qdm provided files.

On all platforms, the applications were compiled with gcc 6.3.0 with -O3 flag. The machines were running Linux version 4.9.0-9. HPL, LAMMPS and NwChem were compiled against Open MPI 3.1.4. Finally, all platform cores were used during all the experiments (28 on CHIFFLET and 64 on YETI) while hyperthreading was disabled.

Regarding DUF, we empirically set the measurement interval to 200 ms. A smaller interval leads to an overhead, while a larger interval may miss some changes in the application behavior and eventually lead to an overhead. We set the tolerated slowdown to 0, 5, 10 and 20% to study how DUF is

[1] commit `aa2b88578`
[2] commit `67f5237ab`

able to respect the tolerated slowdown and its impact on power and energy consumption. We also consider a measurement error of 2% when measuring the flop/s. In other words, as long as the difference between two iterations of DUF is less than 2%, it considers that the flop/s are stable.

6.4.3 Experimental results

Figures 6.5 and 6.6 show the impact of dynamically adapting the uncore frequency with DUF compared to the default behavior on two different architectures on the performance, the power consumption and the total (DRAM + Processor) energy consumption. They also show the impact of the best and worst uncore frequency for each application as a reference limit.

6.4.3.1 DUF impact on performance

On CHIFFLET (Figure 6.5a), one can see that the slowdown is respected for all applications but CG when no slowdown is tolerated. In this case, the observed slowdown is 0.15% (recall that the measurement error is 2%), which is too low to explain the reason behind it. One can also see that for some applications, it is not possible to observe a slowdown, even if the worst configuration shows a high slowdown. This is the case for HPL, NwChem, FT, LU and SP. This is mainly due to the fact that the observed flop/s of these applications keep increasing and decreasing, which does not allow DUF to make a decision on several iterations.

On YETI (Figure 6.6a), DUF is able to respect the tolerated slowdown for all applications except LU with 20% tolerated slowdown. In this case, the observed overhead is 0.11%, which is once again too small to be able to understand the reason behind it.

As a conclusion, DUF is able to respect the tolerated slowdown for 20 out of the 22 tested cases.

6.4.3.2 DUF impact on power consumption

Figures 6.5b and 6.6b show the impact of DUF on the power consumed by CHIFFLET and YETI. They both show that as the slowdown increases, the power savings increase as well for all applications but EP. This is because, as explained in Section 6.4, the uncore frequency has no impact on EP. As a consequence, DUF is able to set the uncore frequency to the minimum available value after few iterations. EP savings reach 16.55% on CHIFFLET and 15.7% on YETI. Regarding the other applications, for most of them, even a small slowdown allows for power savings. For instance, on CHIFFLET, a slowdown of 0.58% on BT allow for power savings of almost 7.94% when the user-defined slowdown is 5%. On YETI, the best power savings when no slowdown is tolerated reach 6.2% for NwChem. A slightly higher slowdown allows for even larger power savings: with 2.98% slowdown, the power savings for NwChem reach 7.31% on YETI. Finally, a higher but still acceptable slowdown can have

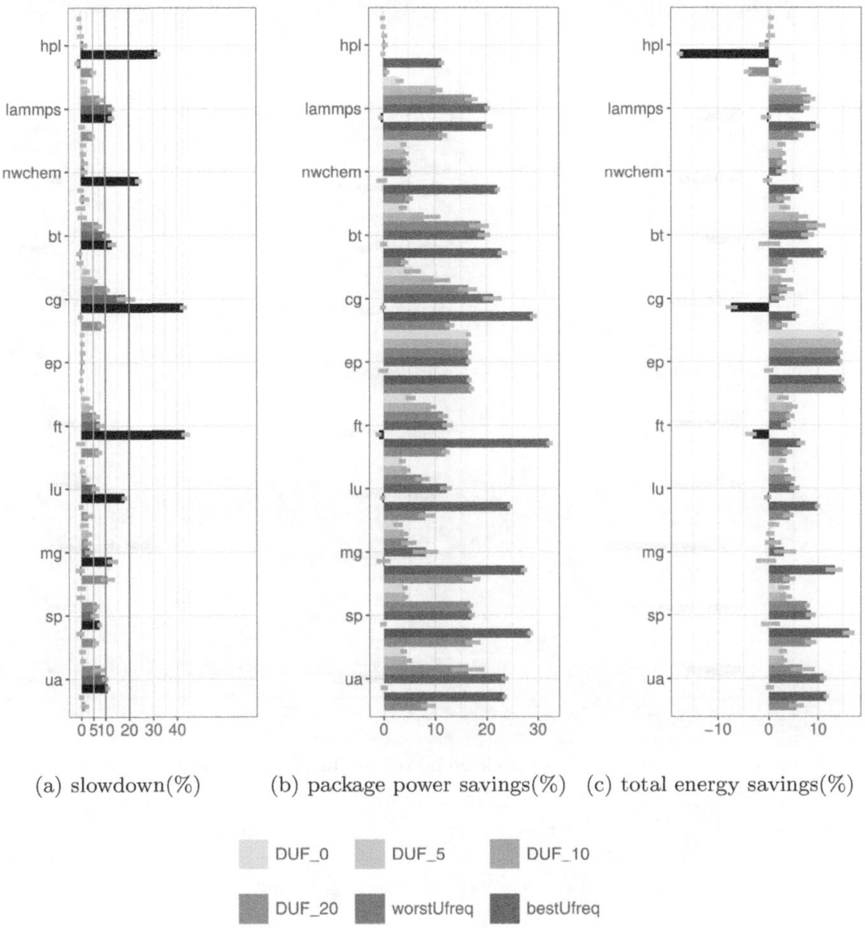

Figure 6.5: DUF impact on performance, power and energy consumption on CHIFFLET.

a large benefit. For instance, on CHIFFLET, a slowdown of 6.3% for SP leads to 16.8% of power savings.

6.4.3.3 DUF impact on energy consumption

Figures 6.5c and 6.6c show the impact of DUF on total (DRAM + sockets) power consumption on CHIFFLET and YETI. Overall, on all platforms, DUF manages to provide energy savings for most applications.

Except for EP, for most applications, the best observed energy savings are reached when the tolerated slowdown is 10% (although the observed slowdown is most of the times around 5 and 7%). For instance, on CHIFFLET, DUF manages to reach 9.7% energy savings with a slowdown of 7.14%, while UA

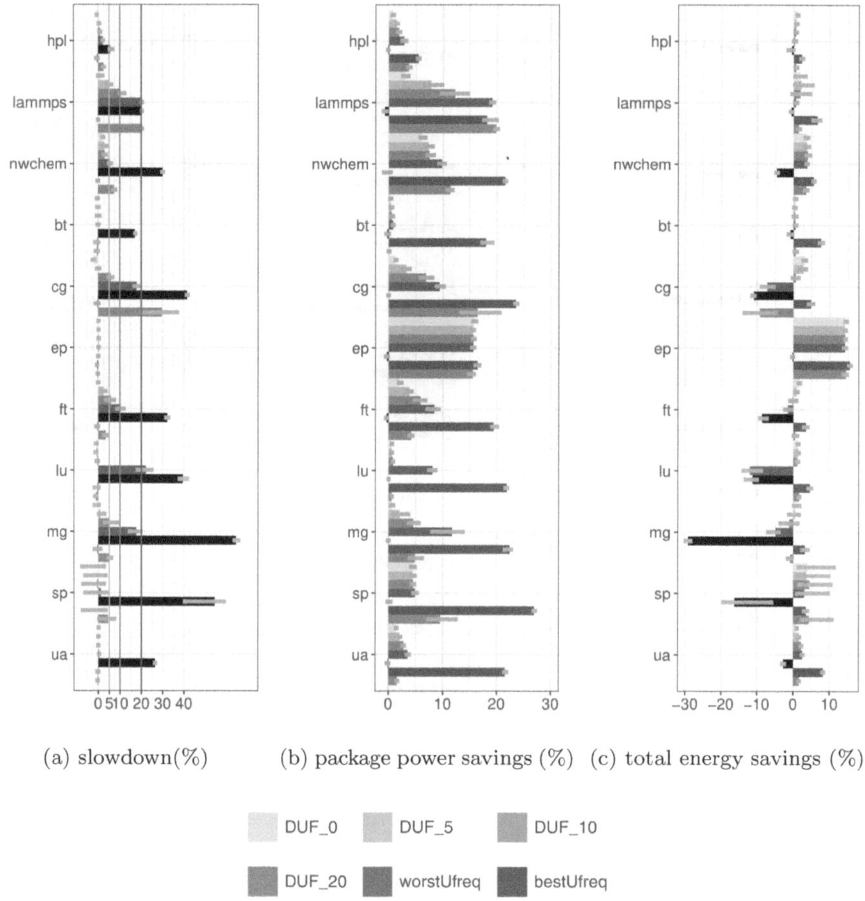

(a) slowdown(%) (b) package power savings (%) (c) total energy savings (%)

Figure 6.6: DUF impact on performance, power and energy consumption on YETI.

reaches 11.1% energy savings with a slowdown of 9.73%. In some cases, an increase in energy efficiency is observed when the tolerated slowdown is 20%. This is the case for MG, LU, SP and UA.

On YETI, DUF manages to provide energy savings but not as much as CHIFFLET. It is able to provide more than 5% energy savings for EP only. Finally, as expected, a high overhead (20%) leads to an increase in energy consumption for most applications.

6.4.4 Improving performance with uncore frequency

As shown in Figure 6.4, it is possible to improve performance with uncore frequency for HPL. However, HPL is an application whose power consumption

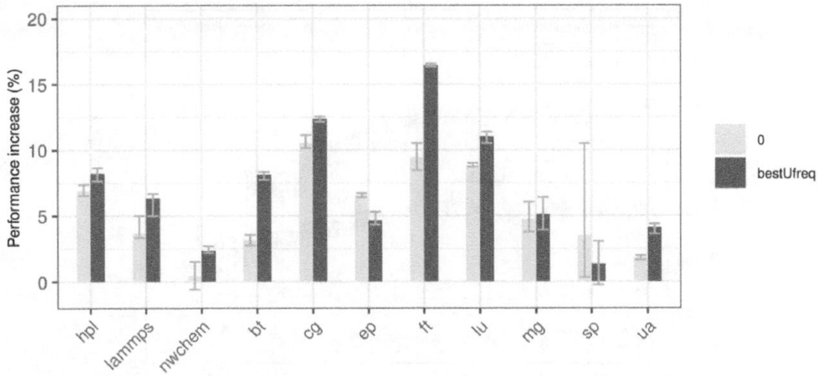

Figure 6.7: Performance increase when using DUF under power capping on YETI.

reaches TDP and is steady throughout the execution. As a consequence, uncore frequency introduces a leverage as decreasing it reduces the power consumption and leaves more budget to the core frequency. In this section, we study how the other applications behave. Most of these applications consume less than TDP or reach it for a short period of time. In order to simulate the behavior observed in Figure 6.4, we use power capping so that all applications can reach the allowed power budget.

Note that the experiment was run on YETI only because it is the only platform that provides power capping. We set the powercap to 100 W for all applications except EP (98 W), NwChem (90 W) and SP (80 W) because these applications have a lower power consumption under normal configuration.

Figure 6.7 depicts the performance improvements under power capping constraints for all applications when no slowdown is tolerated by the user. The results show that for all applications, DUF manages to improve performance. The maximum observed improvement is for CG, with 10.53%.

As explained, this behavior is due to core frequency. For instance, under power capping, we observed that for FT, on average, the default UFS (uncore frequency scaling) sets the uncore frequency to 2.21 GHz, while the core frequency varies between 1.6 and 1.8 GHz. However, DUF sets the uncore frequency to 1.79 GHz, and the core frequency hence reaches 2.02 GHz.

These experiments show that uncore frequency can be used as a leverage to improve performance. However, we set the power capping statically by looking at the average power consumption. In the next section, we present how we can go even further by adapting the power budget to the application needs.

6.5 Power Capping

Power management at the application level became progressively more advanced with the increasing capabilities of frequency and power regulating

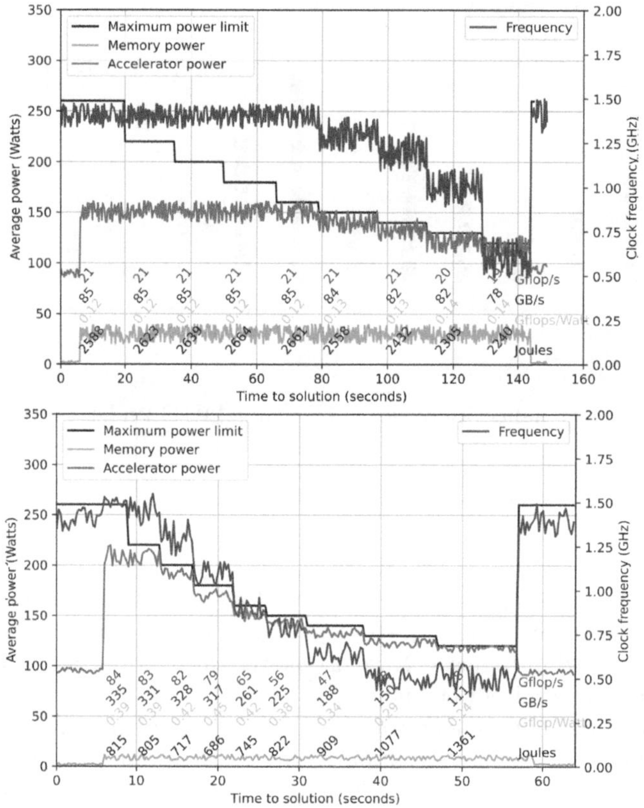

Figure 6.8: Regulating core power capping while simultaneously monitoring energy, bandwidth, and Gflop/s of bandwidth-bound code on Intel Knights Landing to influence the overall power draw in watts. Two memory affinity modes were tested: DDR4 (top) and MCD (bottom).

hardware. This section provides experimental results from power capping and modeling on modern hardware, using results from [12]. Figure 6.8 shows various outcomes of controlling the power draw on Intel Knights Landing (KNL) accelerator. This particular device had a general purpose instruction set even capable of running as a self-hosted processor. However, the figure shows the accelerator mode, which isolated most of the operating system noise to the host CPU. The primary feature of modern power regulation is the model of energy consumption embedded inside the hardware that affects the various compute units and aspects of the chip operation to achieve the user-prescribed power consumption level. The black lines show the stepping down of the power cap that eventually causes the core frequency to drop to maintain the reduced power draw. Another aspect of the test is how the power cap affects the performance metrics of the monitored code, which was a matrix-vector multiply –

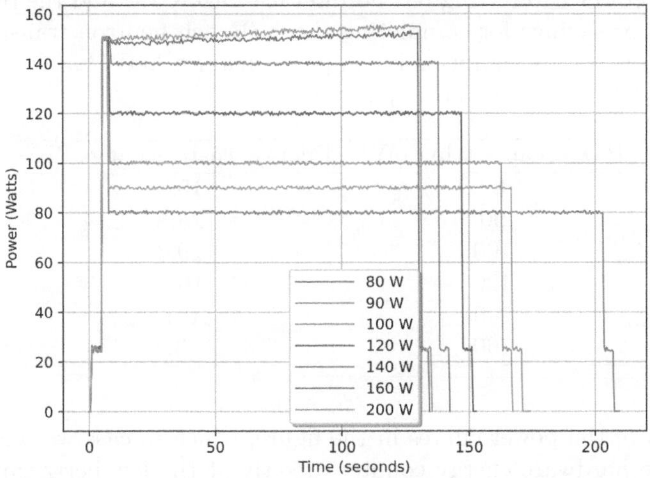

Figure 6.9: A sample calibration sweep of different power capping levels on the Intel Ice Lake CPU to achieve an optimal influence on the maximum power draw when completing the XSBench neutron transport benchmark. The resulting energy waste or savings are shown in Table 6.2.

a bandwidth-bound code. As the power cap lowers, both compute and transfer rates decrease but not at the same rate, because the memory subsystem consumes much less than the compute units. Finally, the figure also shows the energy consumed by the accelerator as the power cap was lowered: the drop in total joules is significant compared to only a modest drop in performance. Thus, the energy-delay product [15] is very favorable on the KNL accelerator for memory-bound codes.

Figure 6.9 shows energy optimization results on a more modern Intel Ice Lake CPU system used throughout this section and in initial reporting [12]. The highest power cap levels are shown as lines for 200 W and 160 W. They show counterintuitive behavior resulting in an inverted level of the reported power draw. This is directly related to the total power envelope allowed for the entire chip and the safety monitoring to maintain TDP. In other words, the user-supplied power cap gets overridden by the thermal sensors that maintain safe temperature of all areas of the CPU die, which might be related to, among other things, the cooling system of the machine room. For the lower power cap levels, we see mostly accurate power draw readings. These kinds of calibration sweeps are an ideal tool for estimating an optimal capping for the targeted energy-delay product [15] or only the total energy consumed. We show in Table 6.2 the relative energy savings (or a waste for the 160 W setting) as reported by the CPU hardware's internal energy consumption model when the runs from Figure 6.9 were done. The energy can also be calculated by approximating the continuous formula $\int_{\text{start}}^{\text{stop}} P(t)dt \approx \sum_{i=0}^{n} P_i \Delta t_i$ or the area

Table 6.2: Power cap setting on the Intel Ice Lake CPU and the resulting energy waste or savings for completing the XSBench neutron transport benchmark. The energy consumption is compared against 200W baseline power cap. See Figure 6.9 for the temporal power draw.

Power cap setting (W)	Energy waste/savings (%)
80	15.9
90	23.8
100	19.8
120	10.1
140	0.1
160	−3.4

under each of the power curves in the figure, which in case was equivalent to reading the hardware energy counter directly at the few hertz sampling rate available to the users.

Figure 6.10 shows more directly the potential trade-offs [12] between two related metrics, time-to-solution and total-energy-consumed, for two drastically different codes: linear solver based on Jacobi iteration for stencil-based discretizations and XSBench code for hash-based lookups of cross-section (XS) energy inside a Monte Carlo method for neutron transport calculations [6, 36].

Both of these codes represent a substantially different performance profile stressing compute and memory subsystems of the CPU in different ways. The solver is a typical memory-bound solver with some floating-point calculation on the transferred data. The XSBench code is much more unstructured and predominantly burdened by integer calculations with less predictable memory traffic that poses a challenge to the data prefetcher logic leading to a stream of irregular addresses requested from the main memory. Consequently, we see a drastically different behavior of the codes reflected in noticeable difference of shapes of the curves in the figure.

The time-to-solution drops sharply for the Jacobi solver when the power cap increases between 80 W and 100 W, while remaining nearly stagnant outside of that range suggesting to retain an aggressive cap for this code. On the other hand, the XSBench code has a much flatter profile practically plateauing above 90 W, leaving the time-to-solution largely unaffected there. The total-energy-consumed is a more complex figure of merit (FOM), as it requires completion of the full test thus causing the total time to vary between the power cap values. And then, the energy is accumulated across the entire runtime taking into the account all the changes in the actual power draw according to the formula shown earlier. This accumulative nature of the total energy and the non-linear influence on time-to-solution that the power cap has both combine to cause a more complex behavior with the curves in the figure indicating minima at 100 W and 90 W for the Jacobi solver and XSBench, respectively.

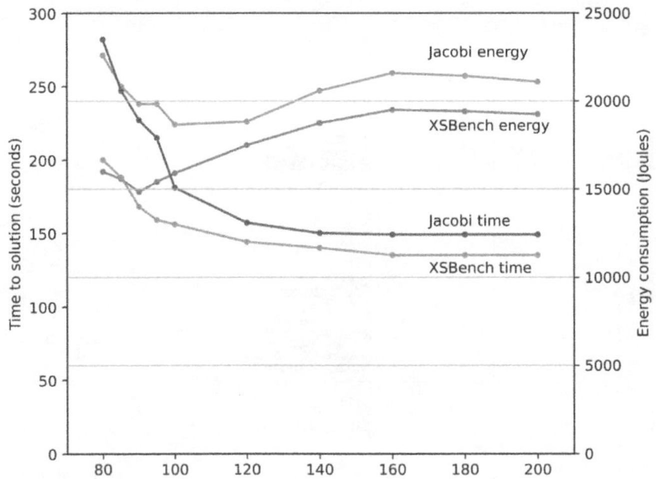

Figure 6.10: Different power capping levels on Intel Ice Lake to influence frequency and power and their corresponding time-to-solution and energy consumption for Jacobi solver and XSBench benchmark codes.

The calibration and time-energy trade-off results presented above serve as a basis to form a fully automated approach. In this chapter, we present two techniques: one based on an online tool which is an extension of DUF, and one that uses a machine learning model for predictions that are then used by a dynamic scheduler to adjust the power configuration at runtime [12].

6.5.1 Dynamic uncore frequency and power capping algorithm

This section introduces Dynamic Uncore Frequency and Power Capping (DUFP) [16], an extension of DUF described in Section 6.4.1. In addition to handling uncore frequency, we want DUF to be able to dynamically adapt the power budget to the application needs. The uncore frequency part remains unchanged, and we will not mention it anymore in this section. The goal here is not to provide energy savings, but rather to provide power savings without degrading the energy consumption.

DUFP algorithm is similar to DUF: it periodically measures the current flop/s. As long as the performance is within the tolerated slowdown (compared to the observed maximum flop/s), it reduces the power budget. As soon as the performance is too impacted, the power budget is increased. Note that just like DUF, DUFP resets the power cap whenever a new phase in the application is detected.

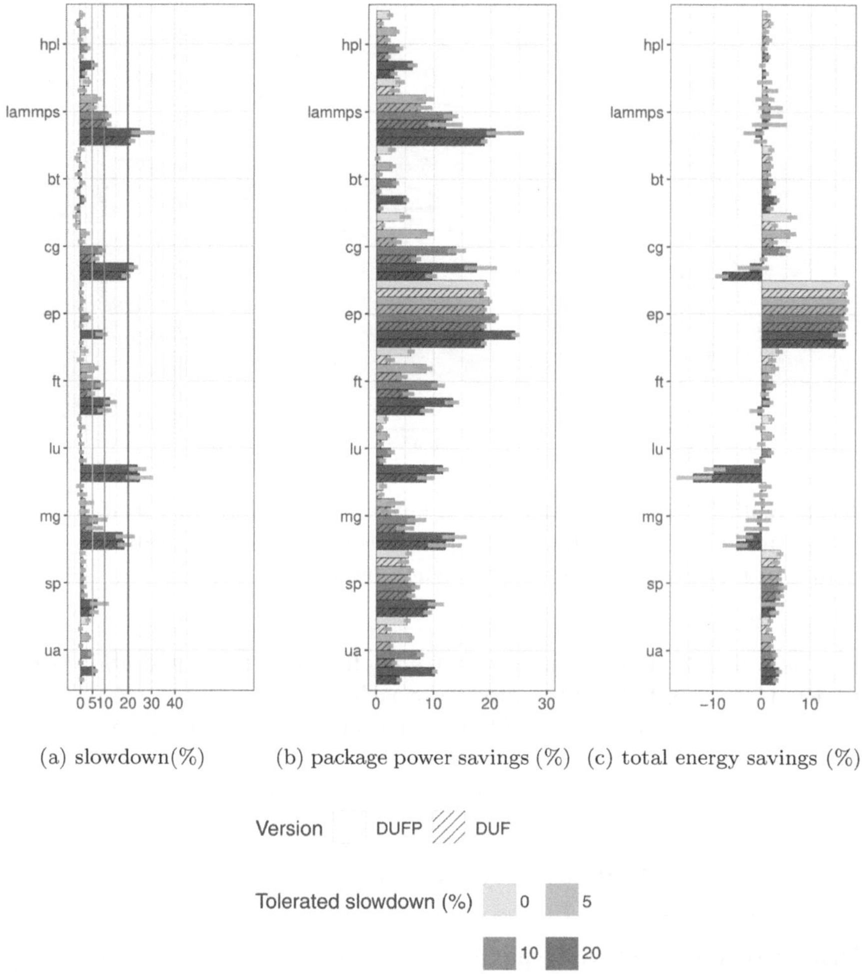

Figure 6.11: DUFP impact on performance, power and energy consumption on YETI.

There is one situation where the decisions for uncore frequency and power capping diverge: since the uncore frequency impacts memory bandwidth, DUF may make decisions based on the behavior of the memory bandwidth. DUFP, on the other hand, does not consider memory bandwidth unless an application has a high compute intensity.

6.5.1.1 Experimental results

Figure 6.11 depicts the impact of dynamic power capping on the performance, the power and the energy consumption compared to the default behavior on

YETI. Note that the experimental setup is the one described in Section 6.4.2. Figure 6.6 also shows the impact of DUF to simplify the comparison. Note also that there may be some differences with Figure 6.6 because of BIOS update between the two rounds of experiments after Spectre and Meltdown discovery.

DUFP impact on performance.

Figure 6.11a shows the impact of DUFP on the performance of the applications. In most cases, DUFP manages to respect the user-defined tolerated slowdown for all applications but the following: LAMMPS at 0 (1.67%), 5 (0.14%) and 20% (3.17%) tolerated slowdown, CG (0.42%) and LU at 20% tolerated slowdown, and UA at 0% slowdown (1.17%). The maximum observed slowdown is 3.17%. The observed slowdown is due to the fact that using a 200 ms interval is too large to detect some power bursts and performance changes. We tried with a 50 ms interval and behaviors that we did not observe at 200 ms. However, a small interval introduces an overhead both on DUF and DUFP. Note that LU shows a similar behavior with DUF and DUFP and the observed overhead is due to the uncore frequency scaling rather than the power capping.

Compared to DUF, using power capping adds an additional leverage, which allows us to have more slowdown for some applications. For instance, HPL and UA show no slowdown regardless of the user-tolerated slowdown with DUF, while DUFP allows for a slowdown.

Overall, DUFP manages to respect the tolerated slowdown for 34 out of the 40 tested configurations.

DUFP impact on power consumption.

Figure 6.11b shows the impact of DUFP on the power consumption. Overall, DUFP manages to show power savings for all applications with a maximum of 24.27% for EP. Compared to DUF, DUFP manages to provide more savings in all configurations with a maximum of 7.90% for CG at 20% tolerated slowdown. It is also able to provide power savings for BT where DUF could not.

Overall, using dynamic power capping in addition to uncore frequency scaling allows for more power savings than uncore frequency scaling by itself.

DUFP impact on energy consumption.

Figure 6.11c shows the impact of DUFP on the processor + DRAM energy consumption. Recall that the goal of DUFP is not to save energy, but rather to decrease power consumption without an additional energy cost.

The results show that DUFP is actually able to provide energy savings for most configurations except for LAMMPS, CG, LU and MG at 20% tolerated slowdown. This is an expected behavior since for these three applications, the slowdown is high compared to the power savings. For most configurations, DUFP is also able to provide even more energy savings than DUF.

6.5.2 Using machine learning to set power capping

The most crucial aspect of such a scheduling system is the fact that all of its components, including the application under control, use the same hardware competing for its resources. Thus, minimizing the overheads and maximizing the accuracy are two essential design constraints without causing undue interference into the application execution with potentially settling for a less exact but more lightweight solution. Accordingly, we choose one of the less complex machine learning models from the *sktime* (scientific kit for time prediction) that offer both explainability and customization enabling downscaling its capabilities to meet the acceptable level of overhead. One such model is random forest regression [5], whose decision trees can be selected based on their size to customize the cost of their traversal during the prediction phase, while choosing the acceptable misprediction error. We trained the model using the runs of the two applications presented above under a number of power settings, and we used the performance counter metrics to arrive at the classification labels that approximated the state of the application and the hardware giving us a decision frontier to pick the next setting from [12]. Other machine learning models were used as well, mostly employing what are now considered traditional methods such as XGBoost, Support Vector Regression, and Random Forest variants to ease the adaptation to a specific set of applications with a three-step process: train, predict, and regulate [18, 34].

Based on the experiments presented so far that indicated the behavior of scientific codes in response to power limit setting, an online scheduling model was created [12] to regulate the power cap in response to the temporal code characteristic and an optimal hardware configuration. Figure 6.12 shows the dynamic scheduling effects based on that model for the two codes analyzed in this section. The core of the model is based on labels (shown to the right in the figure) that are generated dynamically in response to the codes behavior to categorize the code under introspection as either more compute-bound or memory-bound and thus bandwidth limited with a potential to higher energy savings. The labels are decided based on the estimates of intensity of computational power R_{comp}, defined as the ratio of the current package (or core) power and the limiting power of the package:

$$R_{\text{comp}} \stackrel{\text{def}}{=} \frac{P_{\text{pkg}}}{P_{\text{pkg limit}}}. \tag{6.1}$$

Similarly, the memory usage intensity is based on the ratio of power readings for the memory subsystem:

$$R_{\text{mem}} \stackrel{\text{def}}{=} \frac{P_{\text{mem}}}{P_{\text{mem max}}}. \tag{6.2}$$

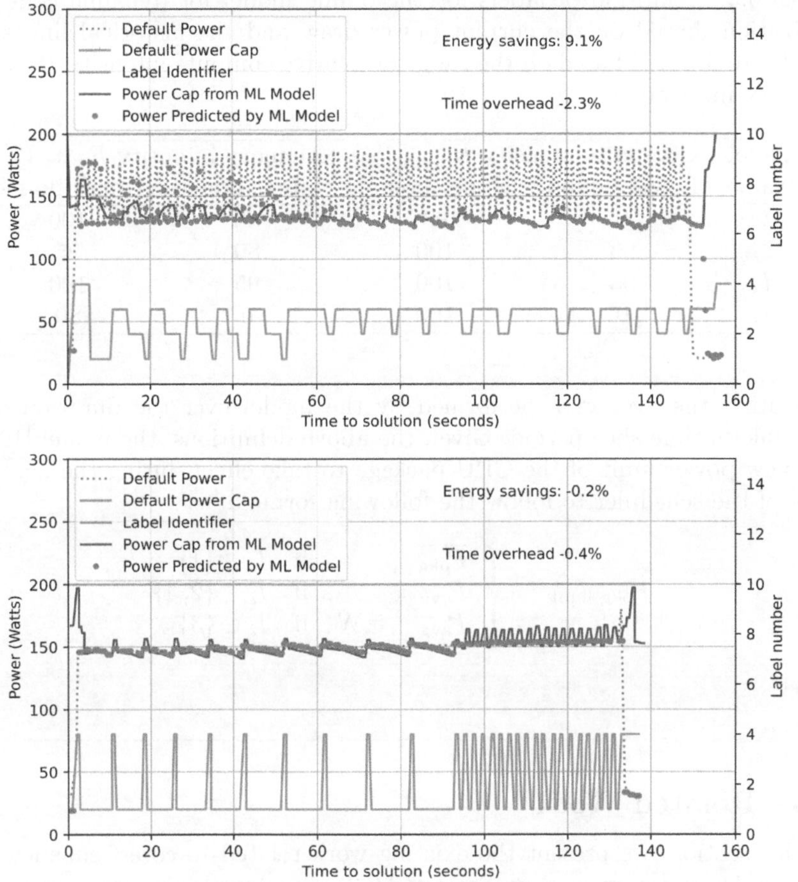

Figure 6.12: Dynamic power capping using a 4-label performance model and corresponding energy consumption for Jacobi solver (top) and XSBench (bottom) codes on Intel Ice Lake processor.

Table 6.3 shows how these relative intensity estimates affect the labeling of the code with labels' number representing the intensity level. The predictive component of the model [12] uses the current power average:

$$P_{\text{avg}} \stackrel{\text{def}}{=} (\overline{P_{\text{pkg}}} - \overline{P_{\text{mem}}})\frac{P_{\text{mem max}}}{\overline{P_{\text{mem}}}} + \overline{P_{\text{mem}}}. \tag{6.3}$$

The averaging uses both the package and memory power and scales the difference of both by the power ratio of memory power readings or simply the memory usage intensity R_{mem}. Note the lines over the power variables

Table 6.3: Definition of labels for the online model for dynamic code categorization, based on the current power draw and the historical limits. The labels are decided based on the range of relative compute intensity R_{comp} and memory intensity R_{mem}.

Label	Compute intensity R_{comp} (%)		Memory intensity R_{mem} (%)	
Name	Lower bound	Upper bound	Lower bound	Upper bound
l_1	0	93	80	100
l_2	93	100	80	95
l_3	93	100	95	100
l_4	0	100	0	80

indicating the averaging performed by the model over the duration of the scheduler's time slice period. Given the above definitions, the model [12] sets the new power limit of the CPU package to take effect during the next time slice of the scheduler to follow the following formula:

$$P_{\text{pkg limit}} \overset{\text{def}}{=} \begin{cases} \overline{P_{\text{pkg}}} & \text{if} \quad l_i \in \{1\} \\ P_{\text{avg}} & \text{if} \quad l_i \in \{2, 4\} \\ P_{\text{avg}} - 5\text{ W} & \text{if} \quad l_i \in \{3\} \end{cases} . \tag{6.4}$$

6.6 Related Work

In this section, we present the existing work related to core frequency (Section 6.6.1), uncore frequency (Section 6.6.2), and dynamic power capping (Section 6.6.3).

6.6.1 Core frequency scaling

The split of the CPU modules into the compute cores and management side called uncore [19] precipitated the need for a set of dedicated methods for adjusting frequency and thus regulating performance, power draw, and energy consumption of the various compute units inside an ever-increasing set of cores. The early approaches using DVFS did not have to make this kind of trade-offs focusing on a specific hardware module [22, 24, 31]. Introduction of RAPL [9] made the split easier to manage, by allowing the software to regulate each package domain independently of all the others and consequently allowing the precise adaptation of the hardware operation to the application-specific workload. Many related efforts in regulating the core frequency may be summarized in focusing on energy savings with potential effects on performance that is subject to many factors at the user level to system-wide considerations [40].

6.6.2 Uncore frequency scaling

Recently, uncore frequency scaling has become a new target for optimization purposes. Some studies use machine learning techniques [3, 39] to find the best uncore frequency to apply throughout the execution, while others combine it to CPU frequency [30, 35], and then later use machine learning to improve the decisions [14]. Unlike DUF, which makes all the decisions at runtime, these approaches first study the applications and then provide a decision.

PowerScavenger [13] is the closest to DUF. It regulates the uncore frequency based on the observed DRAM, the consumed memory power, and the IPC, to avoid any slowdown. We implemented our own version of PowerScavenger and compared it to DUF [1]. We showed that although PowerScavenger is able to provide power and energy savings, by allowing the user to define the tolerated slowdown, DUF is able to provide more savings.

More recently, studies like [8] propose to combine CPU and uncore frequency scaling in the context of the EAR library to improve power consumption. They use the total consumption of the nodes, rather than memory and processors as provided by counters.

6.6.3 Dynamic power capping

Very few studies rely on dynamic power capping to reduce power consumption. Another effort [38] is the closest to DUFP, since it provides the best uncore frequency and power cap to improve energy consumption. It is based on reinforcement learning, while DUFP is an online tool.

Most of the tools that provide a power capping budget are based on models [7, 32] or code instrumentation [21]. Note that one of the models [7] targets memory-intensive applications and could be used in DUFP for memory-intensive phases.

More recently, research switched to designing micro-benchmarks and then use machine learning models to combine power capping and uncore frequency scaling to improve performance and power consumption [38].

6.7 Conclusion

In this chapter, we presented a number of methods for scheduling hardware components and their operation with respect to frequency, voltage, and power caps. The presented algorithms and implementations were devised to address the drastic changes in hardware design, responding to an increase focus on energy consumption and power draw of large datacenters. No static schedule is possible for even a narrowly constrained set of applications drawn from high-performance and large-scale workloads. Consequently, we show how dynamic adjustments allow to scale frequency and cap power while gaining on energy savings with only limited impact on performance. Both parametric and data-based models were used to accumulate the knowledge of the application behavior. We expect this trend of dynamic modeling of performance and adjusting the hardware's low-level setting to continue in the near future.

6.8 References

[1] E. André, R. Dulong, A. Guermouche, and F. Trahay. DUF: Dynamic uncore frequency scaling to reduce power consumption. *Concurrency and Computation: Practice and Experience*, 34(3):e6580, 2022.

[2] D. Bailey, E. Barszcz, J. Barton, D. Browning, R. Carter, L. Dagum, R. Fatoohi, P. Frederickson, T. Lasinski, R. Schreiber, H. Simon, V. Venkatakrishnan, and S. Weeratunga. The NAS parallel benchmarks. *International Journal of Supercomputing Applications*, 5(3):63–73, Sept. 1991.

[3] S. A. Bekele, M. Balakrishnan, and A. Kumar. ML guided energy-performance trade-off estimation for uncore frequency scaling. *Spring Simulation Conference (SpringSim)*, pages 1–12, April 2019.

[4] R. Bolze, F. Cappello, E. Caron, M. Daydé, F. Desprez, E. Jeannot, Y. Jégou, S. Lanteri, J. Leduc, N. Melab, G. Mornet, R. Namyst, P. Primet, B. Quetier, O. Richard, E.-G. Talbi, and I. Touche. Grid'5000: A large scale and highly reconfigurable experimental grid testbed. *International Journal of High Performance Computing Applications*, 20(4):481–494, 2006.

[5] L. Breiman. Random forests. *Machine Learning*, 45(1):5–32, 2001.

[6] F. B. Brown. New hash-based energy lookup algorithm for Monte Carlo codes. *Trans. Am. Nucl. Soc.*, 111:659–662, 2014.

[7] S. Cerf, R. Bleuse, V. Reis, S. Perarnau, and E. Rutten. Sustaining Performance While Reducing Energy Consumption: A Control Theory Approach. In *EURO-PAR 2021 - 27th International European Conference on Parallel and Distributed Computing*, volume 12820 of *Euro-Par*, pages 334–349, Lisbon, Portugal, Aug. 2021. Springer.

[8] J. Corbalan, O. Vidal, L. Alonso, and J. Aneas. Explicit uncore frequency scaling for energy optimisation policies with ear in intel architectures. In *2021 IEEE International Conference on Cluster Computing (CLUSTER)*, pages 572–581, 2021.

[9] H. David, E. Gorbatov, U. R. Hanebutte, R. Khanna, and C. Le. RAPL: memory power estimation and capping. In *Proceedings of the 16th ACM/IEEE international symposium on Low power electronics and design*, pages 189–194, 2010. DOI:10.1145/1840845.1840883.

[10] J. J. Dongarra, P. Luszczek, and A. Petitet. The LINPACK benchmark: Past, present and future. *Concurrency Computat.: Pract. Exper.*, 15(9):803–820, August 10 2003. `http://dx.doi.org/10.1002/cpe.728`.

[11] D. Edelman, S. Samsi, J. McDonald, A. Michaleas, and V. Gadepally. An analysis of energy requirement for computer vision algorithms. In *2023 IEEE High Performance Extreme Computing Conference (HPEC)*, pages 1–7, 2023.

[12] J. Gebert, D. Barry, and J. Dongarra. PowerMan: online power capping by computationally informed machine learning. In *SC22 Proceedings, Poster Session, The International Conference for High Performance Computing, Networking, Storage, and Analysis*, Dallas, TX, 2022.

[13] N. Gholkar, F. Mueller, and B. Rountree. Uncore power scavenger: a runtime for uncore power conservation on HPC systems. *International Conference for High Performance Computing, Networking, Storage and Analysis (SC)*, pages 27:1–27:23, 2019.

[14] A. Gocht, R. Schöne, and M. Bielert. Q-learning inspired self-tuning for energy efficiency in hpc. *International Conference on High Performance Computing Simulation (HPCS)*, pages 344–347, 2019.

[15] R. Gonzalez and M. Horowitz. Energy dissipation in general purpose microprocessors. *IEEE Journal of Solid-State Circuits*, 31(9):1277–1284, Sept. 1996.

[16] A. Guermouche. Combining uncore frequency and dynamic power capping to improve power savings. In *2022 IEEE International Parallel and Distributed Processing Symposium Workshops (IPDPSW)*, pages 1028–1037, 2022.

[17] Y. Guo, D. Cao, X. Xin, Y. Zhang, and J. Yang. Uncore encore: Covert channels exploiting uncore frequency scaling. In *56th Annual IEEE/ACM International Symposium on Microarchitecture (MICRO '23)*, Toronto, ON, Canada, October 28–November 01 2023. ACM, New York, NY, USA. `https://doi.org/10.1145/3613424.3614259`.

[18] M. Hao, W. Zhang, Y. Wang, G. Lu, F. Wang, and A. V. Vasilakos. Fine-grained powercap allocation for power-constrained systems based on multi-objective machine learning. *IEEE Transactions on Parallel and Distributed Systems*, 32(7):1789–1801, 2020.

[19] D. L. Hill, D. Bachand, S. Bilgin, R. Greiner, P. Hammarlund, T. Huff, S. Kulick, and R. Safranek. The uncore: A modular approach to feeding the high-performance cores. *Intel Technology Journal*, 14(3), 2010.

[20] J. Hofmann, D. Fey, J. Eitzinger, G. Hager, and G. Wellein. Analysis of Intel's Haswell microarchitecture using the ECM model and microbenchmarks. In F. Hannig, J. M. P. Cardoso, T. Pionteck, D. Fey, W. Schröder-Preikschat, and J. Teich, editors, *Architecture of Computing Systems – ARCS 2016*, pages 210–222, Cham, 2016. Springer International Publishing.

[21] C. Imes, H. Zhang, K. Zhao, and H. Hoffmann. Copper: Soft real-time application performance using hardware power capping. In *2019 IEEE International Conference on Autonomic Computing (ICAC)*, pages 31–41, 2019.

[22] W. Kim, M. S. Gupta, G.-Y. Wei, and D. Brooks. System level analysis of fast, per-core DVFS using on-chip switching regulators. In *2008 IEEE 14th International Symposium on High Performance Computer Architecture*, pages 123–134. IEEE, 2008.

[23] B. Li, R. Arora, S. Samsi, T. Patel, W. Arcand, D. Bestor, C. Byun, R. B. Roy, B. Bergeron, J. T. Holodnak, M. Houle, M. Hubbell, M. Jones, J. Kepner, A. Klein, P. Michaleas, J. McDonald, L. Milechin, J. Mullen, A. Prout, B. Price, A. Reuther, A. Rosa, M. L. Weiss, C. Yee, D. Edelman, A. Vanterpool, A. Cheng, V. Gadepally, and D. Tiwari. AI-Enabling workloads on large-scale GPU-Accelerated system: Characterization, opportunities, and implications. In *Proceedings of IEEE International Symposium on High-Performance Computer Architecture, HPCA 2022, Seoul, South Korea, April 2-6, 2022*, pages 1224–1237. IEEE, 2022.

[24] P. Macken, M. Degrauwe, M. V. Paemel, and H. Oguey. A voltage reduction technique for digital systems. In *1990 37th IEEE International Conference on Solid-State Circuits*, page 238–239. IEEE, 1990.

[25] J. McDonald, B. Li, N. Frey, D. Tiwari, V. Gadepally, and S. Samsi. Great power, great responsibility: Recommendations for reducing energy for training language models. In M. Carpuat, M.-C. de Marneffe, and I. V. Meza Ruiz, editors, *Findings of the Association for Computational Linguistics: NAACL 2022*, pages 1962–1970, Seattle, United States, July 2022. Association for Computational Linguistics.

[26] A. Petitet, R. C. Whaley, J. Dongarra, and A. Cleary. HPL – a portable implementation of the high-performance Linpack benchmark for distributed-memory computers, September 2000. `http://www.netlib.org/benchmark/hpl`.

[27] S. Plimpton. Fast parallel algorithms for short-range molecular dynamics. *Journal of computational physics*, 117(1):1–19, 1995.

[28] R. B. Roy, T. Patel, V. Gadepally, and D. Tiwari. BLISS: auto-tuning complex applications using a pool of diverse lightweight learning models. In *Proceedings of the 42nd ACM SIGPLAN International Conference on Programming Language Design and Implementation (PLDI 2021)*, page 1280–1295. Association for Computing Machinery, New York, NY, USA, 2021.

[29] S. Samsi, D. Zhao, J. McDonald, B. Li, A. Michaleas, M. Jones, W. Bergeron, J. Kepner, D. Tiwari, and V. Gadepally. From words to Watts: benchmarking the energy costs of large language model inference. In *2023 IEEE High Performance Extreme Computing Conference (HPEC)*, pages 1–9, 2023.

[30] J. Schuchart, M. Gerndt, P. G. Kjeldsberg, M. Lysaght, D. Horák, L. Říha, A. Gocht, M. Sourouri, M. Kumaraswamy, A. Chowdhury, M. Jahre, K. Diethelm, O. Bouizi, U. S. Mian, J. Kružík, R. Sojka, M. Beseda, V. Kannan, Z. Bendifallah, D. Hackenberg, and W. E. Nagel. The readex formalism for automatic tuning for energy efficiency. *Computing*, 2017. doi: `10.1007/s00607-016-0532-7`.

[31] G. Semeraro, G. Magklis, R. Balasubramonian, D. H. Albonesi, S. Dwarkadas, and M. L. Scott. Energy-efficient processor design using multiple clock domains with dynamic voltage and frequency scaling. In *Proceedings Eighth International Symposium on High Performance Computer Architecture*, page 29–40. IEEE, 2002.

[32] S. Sharma, Z. Lan, X. Wu, and V. Taylor. A dynamic power capping library for hpc applications. In *2021 IEEE International Conference on Cluster Computing (CLUSTER)*, pages 797–798, 2021.

[33] O. S. Simsek, J.-G. Piccinali, and F. M. Ciorba. Increasing energy efficiency of astrophysics simulations through GPU frequency scaling. In *Proceedings of The International Conference for High Performance Computing, Networking, Storage, and Analysis (SC24) Workshops Sustainable Supercomputing*, November 2024.

[34] V. Sundriyal and M. Sosonkina. Runtime energy savings based on machine learning models for multicore applications. *Journal of Computer and Communications*, 10(6), 2022.

[35] V. Sundriyal, M. Sosonkina, B. M. Westheimer, and M. Gordon. Comparisons of core and uncore frequency scaling modes in quantum chemistry application gamess. *High Performance Computing Symposium (HPC)*, 13:1–11, 2018.

[36] J. R. Tramm, A. R. Siegel, T. Islam, and M. Schulz. XSBench – the development and verification of a performance abstraction for Monte Carlo reactor analysis. In *PHYSOR 2014 – The Role of Reactor Physics toward a Sustainable Future*, Kyoto, Japan, 2014.

[37] M. Valiev, E. J. Bylaska, N. Govind, K. Kowalski, T. P. Straatsma, H. J. Van Dam, D. Wang, J. Nieplocha, E. Apra, T. L. Windus, et al. NWChem: a comprehensive and scalable open-source solution for large scale molecular simulations. *Computer Physics Communications*, 181(9):1477–1489, 2010.

[38] Y. Wang, W. Zhang, M. Hao, and Z. Wang. Online power management for multi-cores: A reinforcement learning based approach. *IEEE Transactions on Parallel and Distributed Systems*, pages 1–1, 2021.

[39] J. Won, X. Chen, P. Gratz, J. Hu, and V. Soteriou. Up by their bootstraps: Online learning in artificial neural networks for cmp uncore power management. *International Symposium on High Performance Computer Architecture (HPCA)*, pages 308–319, Feb 2014.

[40] H. Zhang, A. Nukada, and Q. Liao. FCUFS: Core-Level Frequency Tuning for Energy Optimization on Intel Processors. In *2024 IEEE International Conference on Cluster Computing (CLUSTER)*, pages 214–225, Los Alamitos, CA, USA, Sept. 2024. IEEE Computer Society.

[41] D. Zhao, S. Samsi, J. McDonald, B. Li, D. Bestor, M. Jones, D. Tiwari, and V. Gadepally. Sustainable supercomputing for AI: GPU power capping at HPC scale. In *Proceedings of the 2023 ACM Symposium on Cloud Computing*, SoCC '23, page 588–596, New York, NY, USA, 2023. Association for Computing Machinery.

Chapter 7

Energy Minimization in Real-Time Systems: from Worst-Case Guarantees to Statistical Learning

Jonatha Anselmi, Bruno Gaujal, Alain Girault, and Stéphan Plassart[1]

This chapter investigates the following question: How can a server with a variable processing speed schedule jobs with hard deadlines while minimizing its energy consumption? We have addressed several variants of this problem. The common context is that we consider a single processing unit with variable processing speed capabilities (thanks to DVFS), and we consider real-time jobs, each one characterized by its release time, its size, and its relative deadline. We present first the simple case of a deterministic environment, where all the job characteristics are known. Two variants are studied here, of increasing uncertainty: (1) when a finite sequence of jobs must be scheduled offline, and (2) when a possibly infinite sequence of jobs must be scheduled online. Then, we move to the more complex case of a stochastic environment: here, some job characteristics become known only when a job arrives (i.e., its relative deadline) or finishes (i.e., its size), and only when statistical information are known (i.e., the size distribution). Three variants are studied here, again of increasing uncertainty: (1) when there is a single job with known statistics, (2) when there are multiple jobs with known statistics, and (3) when there are multiple jobs with completely unknown statistics. The overall picture is that we progressively study variants of the problem where the available information becomes more and more complex. Throughout this presentation, we present the technical results and we provide the scheduling algorithms; mathematical proofs are only given for new results, since proofs for published results can be

[1]This work has been funded by DynamicEnergy Project EMERGENCE 2025 "Reinforcement Learning for Dynamic Minimization of Energy Consumption in Real Time Embedded Systems."

found in the corresponding papers. The techniques we rely on include dynamic programming, Markov Decision Processes, and reinforcement learning.

7.1 Introduction

The question of minimizing energy while satisfying real-time constraints has been extensively studied in the literature from various perspectives, and it is revisited here through a non-classical lens: How can the server exploit its *a priori* knowledge of job characteristics to optimize performance?

The chapter unfolds progressively, reflecting the increasing complexity of the statistical information available to the scheduler. At the first end of the spectrum, we examine the ideal offline deterministic case, where all job characteristics are fully known (Section 7.3). Next, we address the online deterministic case, where no prior information is available, presenting a fundamentally different challenge (also in Section 7.3). Moving beyond deterministic settings, we explore scenarios where job characteristics are stochastic yet their distribution is known, first for a single job (Section 7.4) and then for multiple jobs (Section 7.5). Finally, at the opposite end of the spectrum, we investigate the case where the server has no explicit knowledge of job distributions and must learn an optimal scheduling strategy from past execution data. This evolving perspective is summarized in Figure 7.1.

This chapter is a synthesis of already published papers, with a few noticeable exceptions where new original work is presented for the first time. In Section 7.4.2, Equation (7.24), which provides the optimal speed profile for an arbitrary convex power function and with bounded speeds, is new. In Section 7.5.3, Proposition 7.5 on the optimality of Earliest Deadline First (EDF) is also new; as far as we know, all previous studies using EDF as their default scheduling policy did so without investigating its optimality. Finally, the entire Section 7.6 is novel. To the best of our knowledge, this is the first attempt to apply reinforcement learning in this energy optimization context, with proven guarantees.

7.2 Server and Job Models

7.2.1 Server model

We consider a processing unit (referred to as a *server* in this chapter) with variable processing speed capabilities. This can model a Dynamic Voltage and Frequency Scaling (DVFS) processor and it is characterized by a set $\mathcal{S}$ of

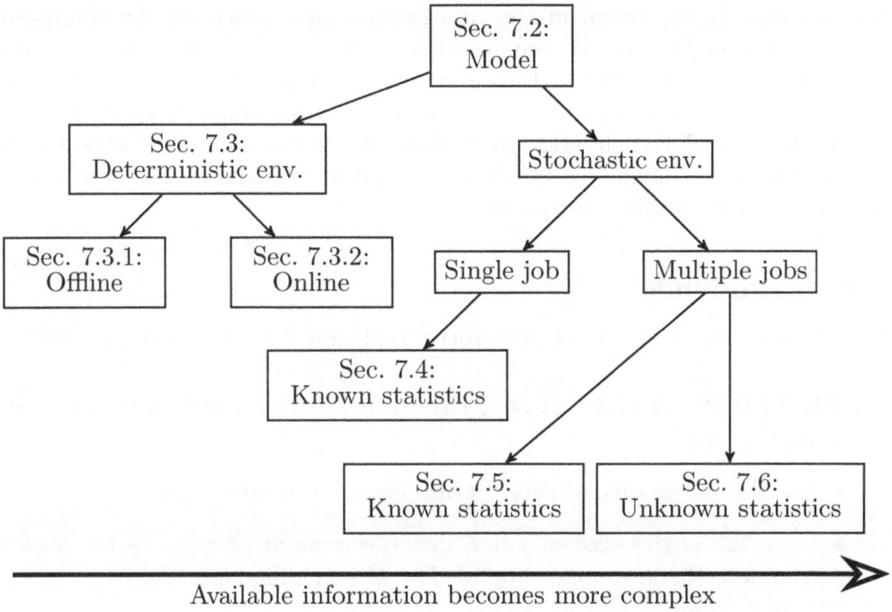

Figure 7.1: Structure of this chapter: The information available to the server becomes more and more complex along the sections.

available speeds with a minimum speed equal to 0 and a maximum speed denoted $s_{\max}$ that can be finite or infinite. For clarity, throughout this work, the term *server* refers to the processing unit responsible for executing jobs, while the term *scheduler* (or *controller*) denotes the control unit that decides which jobs to execute and at what speed.

The power dissipated at any time t by the server running at speed $s(t)$ is denoted $\Phi(s(t))$, while the energy consumed over any period of time $[t_0, t_1]$ is:

$$\mathcal{E}([t_0, t_1]) = \int_{t_0}^{t_1} \Phi(s(t))dt. \tag{7.1}$$

We make no assumption on the function Φ; in particular, we may or may not assume that it is convex, depending on the cases. Indeed, the classical power model is the following [25]: If $s \neq 0$, then $\Phi(s) = \Phi_{leak} + \Phi_{sc} + \Phi_{dyn}(s)$, while if $s = 0$, then $\Phi(s) = \Phi_{idle}$. Here, Φ_{leak} is the leakage power, Φ_{sc} is the short circuit power, $\Phi_{dyn}(s)$ is the dynamic power, proportional to s^3, and finally Φ_{idle} is the power dissipated when the processor is idle. According to this power model, $\Phi(s)$ is a non-convex function of s.

A single-core processor with DVFS can be seen as equivalent to a multi-processor server (without DVFS), where each processor can be either ON or OFF. Specifically, a single-core server operating at speed k corresponds to

k processors being active in the multi-processor setting. In the single-core case, the power function is convex; whereas in the multi-processor case, it is replaced by the concavity of the speed-up function. However, in both cases, power remains convex in the effective speed at which jobs are executed.

In the rest of the chapter, we will use several simplified models for the power function Φ depending on the cases, and we will abstract the resource(s) to one server with multiple speeds.

7.2.2 Job model

We adopt a generic model of preemptive real-time jobs, presented in Def. 7.1.

DEFINITION 7.1 (Job). A **real-time job** J_i is characterized by the triplet (τ_i, w_i, d_i), where:

- $\tau_i \in \mathbb{R}_+$ is the **inter-arrival time** between J_i and J_{i-1};
- $w_i \in \mathbb{R}_+$ is the **size** of job i, i.e., the amount of work to be done to complete the job, upper-bounded by the maximal size $W_{\max}$;
- $d_i \in \mathbb{R}_+$ is the **relative deadline**, upper-bounded by $d_{\max}$.

Here are some important remarks about the job features:

- The size w_i of the job is not to be confused with its Execution Time (ET, in *sec*), nor should the maximal size $W_{\max}$ be confused with the Worst-Case Execution Time (WCET, in *sec*). The size w_i is the number of elementary operations that must be done to complete the job J_i. The relation with its execution time ET_i is $ET_i = w_i/s$, where s is the speed of the processor/server. When the speed of the processor is constant (that is, there is no DVFS), it is common to arbitrarily consider that it is equal to 1. In such a case, the size and the ET are equal, and one can be taken for the other (even though their physical dimensions differ). Here, however, thanks to DVFS, the speed of the processor is changing over time, so the confusion must be avoided.

- This model does not provide directly the release time of the jobs, but actually the *release time* r_i of any job $J_i = (\tau_i, w_i, d_i)$ can be computed iteratively starting from $r_0 = 0$ with the following equation:

$$r_0 = 0 \quad \text{and} \quad \forall i > 0, r_i = r_{i-1} + \tau_i. \tag{7.2}$$

  The release times r_i of the jobs will be used for the formalization in this chapter.

- The *absolute deadline* of job J_i is trivially defined as $D_i = r_i + d_i$ for all i.

This model and notations for jobs and server are quite general and cover most practical examples of embedded systems with hard deadlines. The popular model of real-time systems based on tasks (periodic or sporadic) can be handled by our model by merging all tasks into a single job stream.

All subsequent sections use these conventions and notations up to small local variants that will be specified when needed.

7.3 Deterministic Jobs and Robust Guarantees

This first part is dedicated to the case where the jobs are either fully known (offline) or fully unknown (online). In both cases, no statistical data about the jobs is available.

7.3.1 Offline case

7.3.1.1 Problem statement

Given a finite set of n real-time jobs $\mathcal{J} = \{J_i = (r_i, w_i, d_i)\}_{i \in [1,n]}$ and a time horizon $T = \max_i d_i$, the energy minimization problem is the following:

$$\min_s \quad \mathcal{E}([0,T]) = \int_0^T \Phi(s(t))dt \tag{7.3}$$

$$\text{subject to} \quad \forall 1 \leq i \leq n, \kappa(i, D_i) = w_i \tag{7.4}$$

where D_i is the absolute deadline of job J_i and $\kappa(i,t)$ is the amount of work performed by the processor on job J_i at time t.

7.3.1.2 Critical interval approach

The problem of computing offline schedules to minimize the energy consumption of a finite set of n real-time jobs has been well studied in the literature, starting in 1995 with [26] and [22] where it was independently shown that one can compute the optimal speed schedule with complexity $\mathcal{O}(n^3)$.[2] These two papers assume that the power function $\Phi(s)$ is convex but make no assumption on the number of available speeds.

These algorithms compute the *critical interval* of the set of jobs. Intuitively, a critical interval of a set of jobs $\mathcal{J} = \{J_i\}_{i \in [1,n]}$ is any time interval with the highest load per time unit. In general, there can be several critical intervals for a given set of jobs $\mathcal{J}$. Formally, the critical interval is defined in Def. 7.2:

---

[2] The arithmetic complexity of an algorithm is the number of elementary operations it requires, regardless of the size of their arguments.

DEFINITION 7.2 (Critical interval [26]). Let $\mathcal{J} = \{(r_i, w_i, d_i)\}_{i \in [1,n]}$ be a finite set of real-time jobs, and let T be the finite time horizon. Any time interval $[t_0, t_1] \subseteq [0, T]$ for which the load $\rho([t_0, t_1])$ is maximal is called a *critical interval,* where:

$$\rho([t_0, t_1]) = \frac{\sum_{i \mid [r_i, D_i] \subseteq [t_0, t_1]} w_i}{t_1 - t_0}. \tag{7.5}$$

Yao et al. have proved that, in any optimal speed schedule, all jobs within a critical interval $[t_0, t_1]$ must run at a speed equal to $\rho([t_0, t_1])$ during the entire critical interval. Their theorem assumes that there is no restriction to the set of available speeds. Based on this theorem, they have proposed an algorithm to compute the optimal speed schedule for a given set of real-time jobs $\mathcal{J}$, shown in Algorithm 1. Once a critical interval is found (line 3), and the critical jobs belonging to it are identified (line 4), a temporary vector of speeds is filled from the value $\rho([t_0, t_1])$ (line 5). Then the jobs belonging to the critical interval are removed (line 6) and their characteristics are recomputed (line 7). The **while** loop finishes with the propagation of the temporary speed values computed during this iteration in *temp* with the global vector of speeds s, taking into account the critical interval $[t_0, t_1]$; this is the role of the *CombineSpeeds* function.

The most computationally intensive operation is the search of a critical interval because it requires to check all the possible pairs (r_i, D_j), at a cost of n^2. Moreover, in the worst case, only one job is removed at each iteration of the **while** loop, implying that n such critical intervals must be found. As a result, the overall time complexity of Algorithm 1 is $\mathcal{O}(n^3)$.

Algorithm 1: Computation of the optimal speed schedule based on the critical interval [26].

Input: A finite set of real-time jobs $\mathcal{J} = \{(r_i, w_i, d_i)\}_{i \in [1,n]}$ and the time horizon T.

Output: The optimal speed schedule $s[0, ..., T-1]$.

1 $s[0, ..., T-1] \leftarrow 0$;

2 **while** $\mathcal{J} \neq \emptyset$ **do**

3   Find a critical interval $[t_0, t_1] \subseteq [0, T]$ maximizing $\rho([t_0, t_1])$ computed with Equation (7.5);

4     $\mathcal{J}_{[t_0, t_1]} \leftarrow \{J_i \in \mathcal{J} \mid [r_i, D_i] \subseteq [t_0, t_1]\}$;

5     $temp[t_0, ..., t_1] \leftarrow \rho([t_0, t_1])$;

6     $\mathcal{J} \leftarrow \mathcal{J} \setminus \mathcal{J}_{[t_0, t_1]}$;

7   Collapse the critical interval $[t_0, t_1]$;

8     $s \leftarrow$ CombineSpeeds$(s, temp, t_0, t_1)$;

9 **end**

10 Return s;

The computation of the critical interval and its collapse are illustrated in Figure 7.2.

In 2007, [15] showed that the complexity can be reduced to $\mathcal{O}(n^2 L)$, where L is the nesting level of the set of jobs. In 2017, the complexity has been further reduced to $\mathcal{O}(n^2)$ [18].

In the particular case where the number of available speeds is *finite*, equal to m, [20] gave in 2007 a $\mathcal{O}(m\,n^2)$ algorithm, while [17] proposed in 2005 an $\mathcal{O}(m\,n \log n)$ algorithm. In their most recent work, the same authors showed in 2017 that the complexity can be further reduced to $\mathcal{O}(n \log(\max\{m, n\}))$ [18].

7.3.1.3 Dynamic programming approach

More recently, a dynamic programming approach has been proposed in [13]. Its main contribution is two-fold: on the one hand, to show that computing the optimal speed schedule is not necessarily based on the critical interval, and on the other hand to show that this computation can be linear in the number of jobs to be scheduled. More precisely, the complexity is $\mathcal{O}(nK s_{\max} d_{\max}^2)$ operations, where K is a constant. Besides the complexity gain, one of the main advantages of this approach is that this complexity holds even if the power function Φ is not convex and if speed switching costs are taken into account [13]. In contrast, the previous approaches, presented in Sec. 7.3.1.2, rely heavily on these assumptions. Unlike the previous approach however, it assumes that the job features (r_i, w_i, d_i) are integers, and the set of speeds $\mathcal{S}$ is the finite set of integer speeds, between 0 and $s_{\max}$.

The central idea of the dynamic programming approach of [13] is to encode the current state of the system at time t more compactly than the whole and explicit set of all the jobs active at time t along with their remaining work. Indeed, the scheduling algorithm only needs to know the *cumulative remaining work* at time t, accounting for all the active jobs at time t. It is a staircase function, which we call the *backlog function* and denote $\beta_t(\cdot)$. To compute it, we first introduce in Def. 7.4 the work quantity $A_t(u)$ that arrives at any time t and that must be executed before $t + u$.

Beforehand, we introduce several useful operators in Def. 7.3.

DEFINITION 7.3 (Operators). We define the following operators:

- The time shift operator $\mathbb{T}$ is the shift on the time axis of function f:

$$\forall t \in \mathbb{R}, \quad \mathbb{T}f(t) = f(t+1). \tag{7.6}$$

- $f^+ = \max(f, 0)$ is the positive part of f.

- The unit step function (Heaviside step function) is the discontinuous step function H_t defined for any $t \in \mathbb{R}$ as:

$$\forall u \in \mathbb{R}, \quad H_t(u) = \begin{cases} 0 & \text{if } u < t \\ 1 & \text{if } u \geq t. \end{cases} \tag{7.7}$$

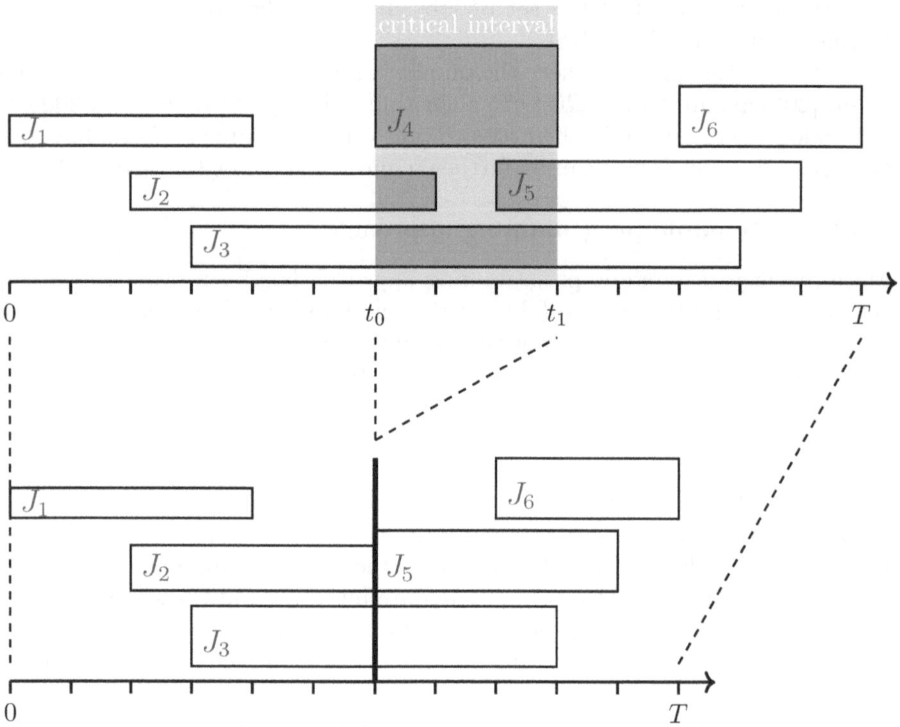

Figure 7.2: The top figure displays a system $\mathcal{J}$ made of six jobs J_1 to J_6, with time horizon T and a critical interval $[t_0, t_1]$. Each job J_i is depicted as a horizontal box starting at r_i, ending at D_i, and with an area equal to w_i. After one loop in Algorithm 1, the critical interval is "collapsed", and all the critical jobs are removed from $\mathcal{J}$ (here only J_4). Moreover, jobs intersecting with the critical interval have their release time updated (here J_5), and/or their relative deadline updated (here J_2, J_3, and J_5). Jobs entirely before the critical interval are unmodified (here J_1), while jobs entirely after the critical interval have only their release time updated (here J_6). The bottom figure displays the same system after the collapse of the critical interval. The former job J_4 has been removed. Jobs J_2, J_3, and J_4 have had their relative deadline shortened. Jobs J_5 and J_6 have had their release time moved earlier. The sizes of the jobs $(w_1, w_2, w_3, w_5, w_6)$ remain unchanged, hence the increased heights of the boxes representing J_2, J_3 and J_5. The duration of the critical interval has been subtracted from the time horizon T.

DEFINITION 7.4 (Arrived work). The amount of work $A_t(u)$ that arrived at time t and must be executed before $t+u$ is:

$$A_t(u) = \sum_{i \in \mathcal{A}_t} w_i \, H_{r_i+d_i}(t+u), \tag{7.8}$$

where $\mathcal{A}_t$ is the set of the jobs released exactly at time t:

$$\mathcal{A}_t := \{J_i = (r_i, w_i, d_i), i \in \mathbb{N} \,|\, r_i = t\}. \tag{7.9}$$

Thanks to Equation (7.8), we can now establish in Equation (7.10) the state change formula when moving from time $t-1$ to time t, using speed $s(u)$ and executing the jobs with the EDF policy in the interval $[t-1, t]$. The backlog β_t at time t is the backlog β_{t-1} at $t-1$ minus the amount of work executed by the processor from $t-1$ to t (which is exactly $\int_{t-1}^{t} s(u)\mathrm{d}u$) plus the new work $A_t(\cdot)$ arriving at t. The max with 0 accounts for the fact that one cannot execute more work during the interval $[t-1, t]$ than what is present at its beginning, and the $\mathbb{T}$ operator performs a shift of the reference time from $t-1$ to t.

PROPOSITION 7.1 (Remaining work [13]). *The remaining work function, or backlog, $\beta_t(\cdot)$ at time $t \in \mathbb{N}$ is:*

$$\beta_t(\cdot) = \mathbb{T}\left[\left(\beta_{t-1}(\cdot) - \int_{t-1}^{t} s(u)\mathrm{d}u\right)^+\right] + A_t(\cdot). \tag{7.10}$$

Figure 7.3 illustrates the definition of $A_t(\cdot)$ and the construction of the backlog $\beta_t(\cdot)$ from $\beta_{t-1}(\cdot)$ and $A_t(\cdot)$.

If $s_t(\beta_t)$ is the work executed by the processor in $[t, t+1]$ in state β_t, then the deadline constraint on the jobs imposes that $s_t(\beta_t)$ must be large enough to at least execute the urgent remaining work at the next time step and cannot exceed the total work present at time t.

Algorithm 2 computes the optimal speed schedule with a dynamic program. Before presenting its pseudo-code, let us provide an informal description.

Let us denote by $\mathcal{E}_t^*(\beta)$ the minimal energy consumption from time t to time T (backward computation), if the backlog at time t is β. Here β is a *variable* that represents *all* possible backlogs at time t. If $\mathcal{E}_t^*(\beta)$ is known *for all* backlogs β at time t, then with a backward induction it is possible to compute the optimal energy consumption $\mathcal{E}_{t-1}^*(\beta)$ at time $t-1$, along with the optimal speed schedule s_{t-1}^*, using the maximum principle:

$$\mathcal{E}_{t-1}^*(\beta) = \min_{s \in \mathcal{F}(\beta)} \left(\Phi(s) + \mathcal{E}_t^*(\mathbb{T}(\beta - s)^+ + A_t)\right) \tag{7.11}$$

$$s_{t-1}^*(\beta) = \operatorname*{arg\,min}_{s \in \mathcal{F}(\beta)} \left(\Phi(s) + \mathcal{E}_t^*(\mathbb{T}(\beta - s)^+ + A_t)\right), \tag{7.12}$$

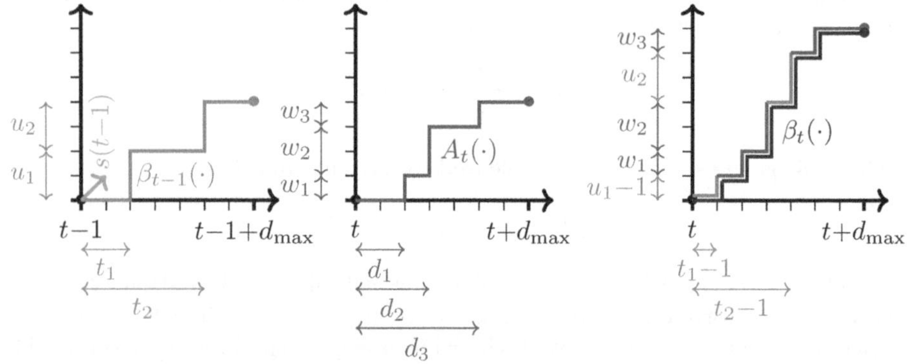

(a) Backlog at $t-1$. (b) Arrival of three jobs. (c) Backlog at t.

Figure 7.3: **(a)** Backlog at $t-1$. The light gray line depicts the remaining work function $\beta_{t-1}(\cdot)$. The constant speed chosen between times $t-1$ and t is $s(t-1)=1$; u_1 stands for $\beta_{t-1}(2)$ and u_2 stands for $\beta_{t-1}(5)-w_{t-1}(2)$. **(b)** Arrival of three new jobs at t: $J_1=(t,1,2)$, $J_2=(t,2,3)$, and $J_3=(t,1,5)$. The dark gray line depicts the corresponding arrival work function $A_t(\cdot)$. **(c)** Resulting backlog at t. The gray line depicts the new remaining work function $\beta_t(\cdot)$, obtained by shifting the time from $t-1$ to t, by executing 1 unit of work (because $s(t-1)=1$), and by incorporating the jobs arrived at t. Above the gray line are shown in light gray the "parts" of $\beta_t(\cdot)$ due to $\beta_{t-1}(\cdot)$ and in dark gray those due to $A_t(\cdot)$.

where $s^*_{t-1}(\beta)$ is the optimal speed at time $t-1$ if the current state at time $t-1$ is β, and $\mathcal{F}(\beta)$ is the set of *admissible* speeds in state β, formally defined as:

$$\mathcal{F}(\beta)=\{s\in\mathcal{S}\mid\beta(d_{\max})\geq s\geq\beta(1)\}. \tag{7.13}$$

When time 0 is reached, the optimal speed schedule has been computed between 0 and $T-1$ *for all possible initial states.* To obtain an optimal speed schedule for the sequence of states $\beta_0,\ldots,\beta_{T-1}$, we just have to return the speeds $s^*_0(\beta_0),\ldots,s^*_{T-1}(\beta_{T-1})$ (see line 25). Note that, because of the "arg min" operator in Equation (7.12), the optimal speed schedule is not necessarily unique.

The cases where it is impossible to schedule the set of jobs without missing a deadline are taken into account by setting the energy function $\mathcal{E}^*_t(\beta')$ to infinity if the state β' is not a possible state, that is, if $\beta'\notin\mathcal{B}$ (see lines 11 and 12), $\mathcal{B}$ being the set of all possible states by definition.

THEOREM 7.1 ([13]). *If the set of jobs is impossible to schedule, then Algorithm 2 outputs "impossible". Otherwise, it outputs an optimal execution in each unit interval that minimizes the total energy consumption.*

Algorithm 2: Dynamic program computing the optimal executed work.

Input: $\{J_i = (r_i, w_i, d_i), i = 1..n\}$ with time horizon T
Output: The optimal speed schedule $\{s_t^*(\beta_t)\}_{t=1...T}$
$\beta_1 \leftarrow (0, \dots, 0)$; {Initial state}
for all $\beta \in \mathcal{B}$ **do** $\mathcal{E}_T^*(\beta) \leftarrow 0$; **end for** {Energy at the horizon}
$t \leftarrow T$; {Start at the horizon}
while $t \geq 1$ **do**
  for all $\beta \in \mathcal{B}$ **do**
    $\mathcal{E}_{t-1}^*(\beta) \leftarrow +\infty$;
    for all $s \in \mathcal{F}(\beta)$ **do**
      $\beta' \leftarrow \mathbb{T}[(\beta - s)^+] + A_t$; {Compute the next state}
      if $\beta' \notin \mathcal{B}$ **then**
        $\mathcal{E}_t^*(\beta') \leftarrow +\infty$; {The next state is unfeasible}
      end if
      if $\mathcal{E}_{t-1}^*(\beta) > \Phi(s) + \mathcal{E}_t^*(\beta')$ **then**
        $\mathcal{E}_{t-1}^*(\beta) \leftarrow \Phi(s) + \mathcal{E}_t^*(\beta')$; {Update the energy in β at $t-1$}
        $s_{t-1}^*(\beta) \leftarrow s$; {Update the optimal speed in β at $t-1$}
      end if
    end for
  end for
  $t \leftarrow t - 1$; {Backward computation}
end while
if $\mathcal{E}_1^*(\beta_1) = +\infty$ **then**
  return "impossible";
else
  return $\{s_t^*(\beta_t)\}_{t=1...T}$;
end if

The time complexity of Algorithm 2 is Kn, where n is the number of jobs and the constant K depends on the maximal speed $s_{\max}$ and the maximal relative deadline $d_{\max}$.

Several improvements have been proposed in [13], first to reduce the time complexity from $\mathcal{O}(n \times \sqrt{d_{\max}}(e\, s_{\max})^{d_{\max}+1})$ to $\mathcal{O}(n \times (s_{\max} W_{\max} d_{\max}^2))$, and second to take into account the non-null time and/or energy switching cost from one speed to another.

7.3.2 Online case

In the online case, we are given a possibly infinite sequence of jobs $\mathcal{J} = \{(r_i, w_i, d_i)\}_i$. The job features (r_i, w_i, d_i) are *unknown to the scheduler.* For simplicity, we assume that they are all integers.

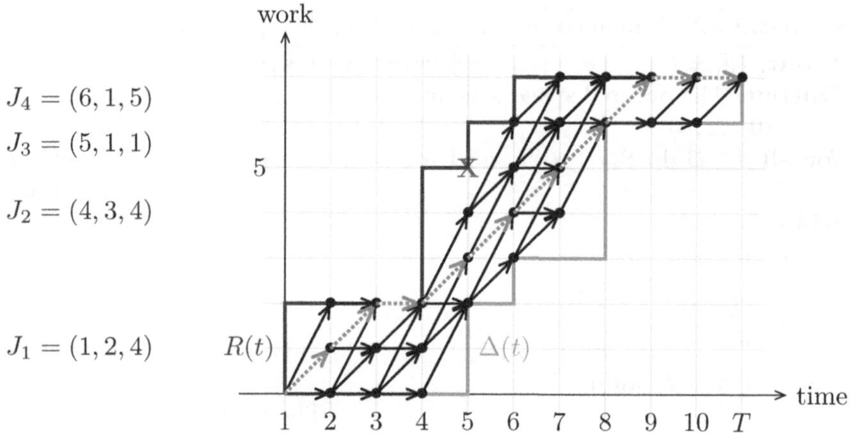

Figure 7.4: Execution of Algorithm 2 with four jobs $\{J_i = (r_i, w_i, d_i)\}_{1=1..4}$ and available speeds $s \in \{0, 1, 2\}$. The cumulative deadlines form the light gray staircase $\Delta(t)$, while the cumulative releases form the dark gray staircase $R(t)$. All the visited states are aggregated as black dots and all the valid state changes as black arrows. The optimal speed schedule computed by Algorithm 2 is shown with gray dotted arrows: It is $(1, 1, 0, 1, 1, 1, 1, 1, 0, 0)$. No state corresponding to the point $(5, 5)$ is visited (marked with the light gray cross), because it is not reachable at maximal speed 2. Finally, speed 0 is not considered between times 5 and 6 because $\beta_5(1) = 1$ for all states at time 5. This results from Job J_3 that arrives at time 5 with a relative deadline of 1. This is not explicitly shown in the figure.

In that case, the performance of a scheduler is measured by two criteria: its *competitive ratio* with respect to the best offline scheduler, and its *feasibility region*. Their study is the goal of the two upcoming sections, Sections 7.3.2.1 and 7.3.2.2.

7.3.2.1 Competitive ratios

Several schedulers have been proposed to execute online jobs with deadlines while minimizing the energy consumption. The most popular ones are Optimal Available (OA) and AVerage Rate (AVR), introduced in [26], and Bansal, Kimbrel, Pruhs (BKP) policy, proposed in [5]. Because no statistical data is available to the scheduler, all these speed policies use only the features of the *active jobs*. The case where the statistics of the job features are known to the scheduler will be presented in Section 7.5.

DEFINITION 7.5 (Optimal Available (OA) [26]). At each time t, the job that has the earliest deadline is executed at speed:

$$s^{(\mathrm{OA})}(t) = \max_{v>t} \left(\frac{\beta_t^{(\mathrm{OA})}(v)}{v - t} \right), \tag{7.14}$$

where $\beta_t^{(OA)}(.)$ is the backlog function at time t under the (OA) speed policy.

DEFINITION 7.6 (AVerage Rate (AVR)). At each time t, the job that has the earliest deadline is executed at speed:

$$s^{(AVR)}(t) = \sum_{i \in \mathcal{R}(t)} \frac{w_i}{d_i}, \tag{7.15}$$

where $\mathcal{R}(t)$ is the set of *released* jobs at time t, i.e., jobs $J_i = (r_i, w_i, d_i)$ such that $r_i \leq t < d_i + r_i$.

The definition of (BKP) policy relies on the concept of contributing work.

DEFINITION 7.7 (Contributing work). For any t, t_1, and t_2 in $\mathbb{R}$ such that $t_1 \leq t \leq t_2$, $u(t, t_1, t_2)$ is the amount of work arrived after t_1 and before t, the deadline of which is less than t_2.

According to Def. 7.7, any job $J_i = (r_i, w_i, d_i)$ contributing to $u(t, t_1, t_2)$ must satisfy $t_1 \leq r_i \leq t$ and $d_i \leq t_2$.

DEFINITION 7.8 (Bansal, Kimbrel, Pruhs policy (BKP)). At each time t, the job that has the earliest deadline is executed at speed:

$$s^{(BKP)}(t) = \max_{t_2 > t} \left\{ \frac{u(t, et - (e-1)t_2, t_2)}{t_2 - t} \right\}. \tag{7.16}$$

The competitive ratio of these three policies have been studied in [5].

THEOREM 7.2 ([5]). *When the power function* Φ *is of the form* $\Phi(s) = s^\alpha$, *with* $\alpha > 1$, *then*

- (OA) *is* α^α *competitive.*
- (AVR) *is* $2^{\alpha-1}\alpha^\alpha$ *competitive.*
- *The competitive ratio of* (BKP) *is at most* $2\left(\frac{\alpha}{\alpha-1}\right)^\alpha e^\alpha$.

It is worth noting that (BKP) has a better competitive ratio than both (OA) and (AVR) as soon as $\alpha \geq 5$. In [5], the authors further show that (BKP) never uses a speed larger than exp(1) times the maximal speed used by the optimal offline policy. This point is actually related to similar bounds for the feasibility region of (BKP) given below.

7.3.2.2 Feasibility regions

It should be noted that none of these three policies account for the server's speed being bounded by a finite maximum speed $s_{\max}$: arbitrarily high speeds must be available to ensure that no deadlines are missed.

In the more realistic case where processor speed is bounded, the performance of any online policy is not uniquely evaluated by its comparison with the best offline policy, but also by the largest set of jobs it can handle. This set is referred to as its *feasibility region* (formalized in Def. 7.10).

Let us denote by $\mathscr{J}[W_{\max}, d_{\max}]$ the set of all possible sequences of jobs whose deadline and size are bounded.

DEFINITION 7.9 (Set of all possible sequences of jobs)**.** This set is denoted $\mathscr{J}[W_{\max}, d_{\max}]$ and equal to:

$$\left\{ \mathcal{J} = \{J_i = (r_i, w_i, d_i)\}_{i\in\mathbb{N}} \middle| \forall t, \sum_{i \mid r_i = t} w_i \leq W_{\max} \wedge \forall i, d_i \leq d_{\max} \right\}. \quad (7.17)$$

In words, a stream of jobs $\mathcal{J} = (J_i)_{i\in\mathbb{N}}$ is in $\mathscr{J}[W_{\max}, d_{\max}]$ if the total amount of work that can arrive instantaneously is upper-bounded by $W_{\max}$, and the relative deadline of any job is bounded by $d_{\max}$.

DEFINITION 7.10 (Speed policy's feasibility)**.** An *online speed policy* π is feasible over an infinite sequence of jobs $\mathcal{J} = \{(r_i, w_i, d_i)\}_{i\in\mathbb{N}}$ if and only if when the processor runs at speed $s^\pi(t)$ smaller than $s_{\max}$ and uses EDF, each job (r_i, w_i, d_i) is executed before its deadline (i.e., $\forall t, \beta_t^\pi(0) = 0$):

$$\pi \text{ is feasible} \iff \left(\sup_{\mathcal{J}\in\mathscr{J}[W_{\max}, d_{\max}]} \sup_t s^\pi(t) \leq s_{\max} \right) \\ \wedge \left(\forall \mathcal{J} \in \mathscr{J}[W_{\max}, d_{\max}], \forall t,\ \beta_t^\pi(0) = 0 \right). \quad (7.18)$$

For policies (OA), (AVR) and (BKP), necessary and sufficient conditions of feasibility were given in [12] and are summarized in the following Table 7.1, where $\mathfrak{h}(n)$ is the n-th harmonic number: $\mathfrak{h}(n) = 1 + \frac{1}{2} + \cdots + \frac{1}{n}$. The proofs are quite involved and are not reported here. For (BKP), we distinguish two cases depending on whether the *decision times* of the scheduler — the instants at which the scheduler is allowed to change the processor speed — are integers or reals (not to be confused with the job features, which are always integers).

Table 7.1: Necessary and sufficient feasibility conditions of the four online speed policies.

Policy	Necessary and sufficient feasibility condition	
Decision times	$\mathbb{N}$	$\mathbb{R}$
(OA)	$s_{\max} \geq W_{\max}(\mathfrak{h}(d_{\max} - 1) + 1)$	
(AVR)	$s_{\max} \geq W_{\max}\mathfrak{h}(d_{\max})$	
(BKP)	$s_{\max} \geq \frac{3}{2}(e-1)W_{\max} \simeq 2.577\, W_{\max}$	$s_{\max} \geq eW_{\max} \simeq 2.718\, W_{\max}$

THEOREM 7.3 ([12]). *The feasibility regions of the three online policies* (OA), (AVR), *and* (BKP) *are:*

For (BKP), the feasibility condition differs depending on whether the scheduler can adjust the server speed at any time (the decision times belong to $\mathbb{R}$) or only at integer points (the decision times belong to $\mathbb{N}$). The respective feasibility regions $\mathcal{F}_{(\mathrm{OA})}, \mathcal{F}_{(\mathrm{AVR})}, \mathcal{F}_{(\mathrm{BKP})}$ are formally compared in Table 7.2.

Table 7.2: Feasibility region comparisons for (OA), (AVR), and (BKP).

Feasibility regions $\mathcal{F}_\pi$	$\mathcal{F}_{(\mathrm{AVR})} \subset \mathcal{F}_{(\mathrm{BKP})}$	$\mathcal{F}_{(\mathrm{OA})} \subset \mathcal{F}_{(\mathrm{BKP})}$	$\mathcal{F}_{(\mathrm{OA})} \subset \mathcal{F}_{(\mathrm{AVR})}$
Integer decision times	$\forall d_{\max} \geq 7$	$\forall d_{\max} \geq 4$	$\forall d_{\max} \in \mathbb{N}$
Real decision times	$\forall d_{\max} \geq 9$	$\forall d_{\max} \geq 4$	$\forall d_{\max} \in \mathbb{N}$

Unlike for (OA) and (AVR), the feasibility region bounds for (BKP) are independent of the maximal deadline $d_{\max}$. This means that the (BKP) feasibility regions do not change when $d_{\max}$ grows, whereas for (OA) and (AVR), the feasibility region decreases to the empty set when $d_{\max}$ increases.

7.4 Single Stochastic Job

In this section, we move to a *stochastic* setting where statistical information is available to the scheduler, and we consider the problem of identifying the optimal processing speed profile for a *single* job. The case of multiple jobs will be addressed in the next section. We observe that even the single-job case is non-trivial and has been used in the literature to develop heuristics for handling multiple jobs arriving stochastically over time [9].

7.4.1 Problem statement

The job size, denoted by W, is a *random variable with cumulative distribution function* $F : F(w) = \mathbb{P}(W \leq w)$, and *finite support* $[0, W_{\max}]$, both known to the processor. Without loss of generality, we assume that the job arrives at time 0 and has a deterministic deadline D. In this section, the power function Φ is assumed to be convex. The goal is to compute a speed profile for the processor that ensures that the job is completed before D while minimizing the expected energy consumption. In mathematical terms, our objective is to find a speed profile $s : [0, W_{\max}] \to [0, s_{\max}]$ that minimizes the mean energy consumption to execute the given job constrained by the deadline D. The mean energy consumption under speed profile s, denoted as $\mathcal{E}(s)$, is defined by Eq. (7.19).

By conditioning on the size ($W = w$) of the task, we have

$$
\begin{aligned}
\mathcal{E}(s) &\stackrel{\text{def}}{=} \int_0^{W_{\max}} \left(\int_0^w \frac{\Phi(s(x))}{s(x)} \mathrm{d}x \right) \mathrm{d}F(w) \\
&= \int_0^{W_{\max}} \frac{\Phi(s(x))}{s(x)} \left(\int_x^{W_{\max}} \mathrm{d}F(w) \right) \mathrm{d}x \\
&= \int_0^{W_{\max}} F^c(x) \frac{\Phi(s(x))}{s(x)} \mathrm{d}x \qquad (7.19)
\end{aligned}
$$

where the second equality follows by changing the order of integration. Therefore, our objective is to solve the following infinite-dimensional optimization problem:

$$
\min_s \int_0^{W_{\max}} \frac{\Phi(s(w))}{s(w)} F^c(w) \, \mathrm{d}w \qquad (7.20)
$$

$$
\text{subject to:} \int_0^{W_{\max}} \frac{\mathrm{d}w}{s(w)} \leq D. \qquad (7.21)
$$

The constraint in Eq. (7.21) states that the job has to be completed before its deadline D. To see this, let $\theta(w)$ denote the time to execute the first w epochs. Then,

$$
\frac{\mathrm{d}\theta(w)}{\mathrm{d}w} = \frac{1}{s(w)}. \qquad (7.22)
$$

By integrating both sides and using the facts that $\theta(0) = 0$ and that $\theta(W_{\max}) \leq D$, we obtain Eq. (7.21).

In the following, we will denote by $Q(s)$ the power per work unit when operating at speed s

$$
Q(s) := \Phi(s) \frac{\mathrm{d}t}{\mathrm{d}w} = \frac{\Phi(s)}{s}. \qquad (7.23)
$$

Since $\Phi(0) = 0$ and Φ is convex, Q is non-decreasing. Clearly, the above optimization problem admits a feasible solution if and only if $s_{\max} \geq W_{\max}/D$, a condition that we assume in the following.

7.4.2 Continuous speeds

The optimization problem (7.20)–(7.21) has been solved in [19] using Jensen inequality when (i) the processor speed is not bounded, and (ii) the power function is of the form $\Phi(s) = Cs^k$, $k > 1$. Here, we give the general form of the solution, which covers the cases where the maximal speed is bounded and power function is "arbitrary". Instead of Jensen inequality, this solution is obtained using the Pontryagin maximum principle as follows. First, let us define the Hamiltonian $h(\theta(w), s(w), \lambda(w), w) = F^c(w)Q(s(w)) - \lambda(w)/s(w)$, where $\lambda(w)$ is the Lagrangian multiplier. The Pontryagin maximum principle says that the optimal solution (s^*, λ^*) satisfies the following conditions:

1. $h(\tau^*(w), s^*(w), \lambda^*(w), w) \leq h(\theta(w), s(w), \lambda(w), w)$ for all $w \leq W_{\max}$.
2. $\lambda^{*\prime}(w) = \lambda^*(w)\frac{\partial(1/s^*(w))}{\partial\tau} + \frac{\partial(F^c(w)Q(s^*(w)))}{\partial\tau}$.

The second condition implies that $\lambda^*(w)$ is a constant denoted λ^*, and the first condition gives the optimal speed under an implicit form.

THEOREM 7.4. *The optimal speed profile satisfies*

$$s^*(w)^2 Q'(s^*(w)) = \frac{\lambda^*}{F^c(w)}, \text{ if } s^*(w) \in [0, s_{\max}]. \tag{7.24}$$

Eq. (7.24) has a unique non-negative solution because $\Phi(s)$ is convex, which implies that $s^2Q'(s)$ is non-decreasing. In the classical case, where $Q(s) = Cs^k$, this gives

$$s^*(w) = K^{-1} F^c(w)^{-1/(k+1)} \wedge s_{\max}, \tag{7.25}$$

where K is a constant such that $\int_0^{W_{\max}} \frac{1}{s^*} = D$, which ensures the validity of the deadline constraint.

Figure 7.5 displays the optimal solution s^* when $k = 2$ and the job size is uniformly distributed over $[0, W_{\max}]$. In that case,

$$\int_0^{W_{\max}} \left(K \left(1 - \frac{w}{W_{\max}}\right)^{1/3} \vee \frac{1}{s_{\max}} \right) dw = D. \tag{7.26}$$

7.4.3 Discrete speeds

We now consider the case where only a finite number of speeds, say $s_1, \ldots, s_N$, are available to the processor. This case has been studied in [1]; we point the reader to this reference for further details and proofs. The role of the scheduler is to choose which speed to activate at any point in time, or equivalently after any amount of completed work. We assume that speed changes are immediate and do not incur any additional energy cost; for instance, ramp-up and/or speed change effects are neglected.

In the setting above, the following structural property holds.

PROPOSITION 7.2 ([1]). *There exists an optimal speed profile that is increasing.*

The previous proposition implies that we can restrict the analysis to increasing speed profiles. The clear advantage of increasing profiles is that they can change at most $N-1$ times, at points, say, $x_1, \ldots, x_{N-1}$. Thus, the optimization problem under investigation boils down to

$$\min_s \sum_{i=1}^N Q(s_i) \int_{x_{i-1}}^{x_i} F^c(w)\, dw \tag{7.27}$$

$$\text{subject to} \sum_{i=1}^N \frac{x_i - x_{i-1}}{s_i} = D, \tag{7.28}$$

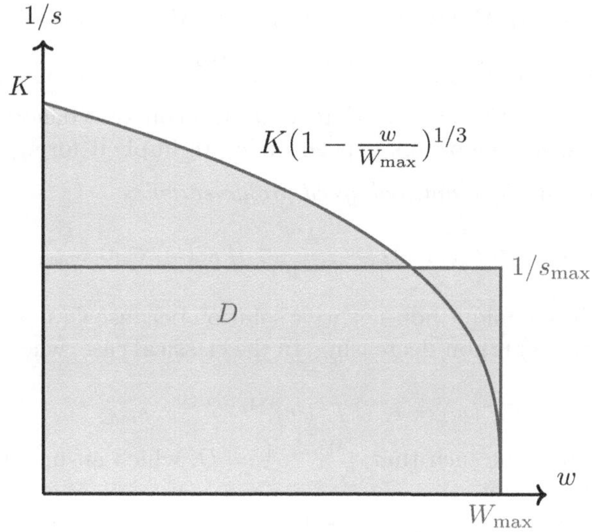

Figure 7.5: Illustration of the optimal speed profile s^* when the job size follows a uniform distribution ($F^c(w) = 1 - \frac{w}{W_{\max}}$). The starting point K is chosen such that the area of the colored zone is D.

which is a convex program. In addition, if F is continuous, then the objective function is differentiable, and an optimizer can be computed efficiently in polynomial time by using, e.g., interior point methods [7]. Otherwise, one can rely on existing algorithms from derivative-free optimization theory [8]. In principle, this analysis provides a satisfactory solution when the number of speeds is finite. However, this approach does not fully exploit the particular structure of the considered problem, because the previously mentioned optimization algorithms are general purpose. This raises the natural question: *Can we do better than general numerical computation via convex programming?*

In the remainder of this section, we assume that F is continuous and that Φ is strictly convex. Within these assumptions, the following two structural properties hold.

PROPOSITION 7.3 ([1]). *Let σ^* be an optimal schedule. Then, (i) σ^* uses a consecutive set of speeds.*
(ii) σ^* uses the maximal speed s_N.

Both properties in Prop. 7.3 can be proven by contradiction and strongly rely on the strict convexity of the power function Φ [1]. For (7.3)(i), it is possible to show that if a speed profile s uses speed s_{i+1} directly after s_{i-1}, then a better speed profile can be constructed if both s_{i-1} and s_{i+1} would be replaced by s_i. Similarly, for (7.3)(ii), one initially assumes that the last speed used in the optimal schedule is s_{N-1} and then shows that it is possible to construct a better speed profile by using s_N as the last speed.

Therefore, if an optimal speed profile uses speed s_i, then it also uses speed s_j for all $j = i, \ldots, N$. Note that an optimal profile may not use the lowest speed s_1, which is intuitive if the deadline is short. These properties, all possessed by an optimal speed profile, can be exploited to design a lower complexity optimal algorithm. This is done in the remainder of this section.

Given $m \in \{1, \ldots, N\}$, let us construct the speed profile $v := v_m : [0, W_{\max}] \to \{s_m, \ldots, s_N\}$ as follows:

$$\forall m \leq i \leq N,\ v(w) = s_i \text{ if and only if } w \in [y_{i-1}, y_i), \tag{7.29}$$

where $y_0 = \cdots = y_{m-1} = 0$, $y_N = W_{\max}$ and the speed change vector point $(y_m, \ldots, y_{N-1}) \in [0, W_{\max}]^{N-m}$ satisfy

$$y_i = (F^c)^{-1}\left(\frac{\Gamma_m}{\Gamma_i} F^c(y_m)\right), \quad i = m+1, \ldots, N-1 \tag{7.30}$$

where $(F^c)^{-1}(y) \stackrel{\text{def}}{=} \sup\{z : F^c(z) = y\}$,

$$\Gamma_i \stackrel{\text{def}}{=} \frac{s_i \Phi(s_{i+1}) - s_{i+1}\Phi(s_i)}{s_{i+1} - s_i} \tag{7.31}$$

and

$$\sum_{i=m}^{N} \frac{y_i - y_{i-1}}{s_i} = D. \tag{7.32}$$

We notice that the structure of the profile v is consistent with the structural properties in Propositions 7.2 and 7.3.

The next proposition gives a property on v.

PROPOSITION 7.4 ([1]). *Assume that F is continuous and Φ is strictly convex. Then, there exists a unique optimizer that solves (7.20)-(7.21). In addition, if such optimizer uses all the speeds $s_1, \ldots, s_N$, then it is given by v with $m = 1$.*

By combining the previous propositions, we can design an algorithm that computes the optimal schedule *even when some speeds are not used in the optimal solution.* This is given in Algorithm 3.

Specifically, we have the following result.

THEOREM 7.5 ([1]). *Assume that F is continuous and Φ is strictly convex. Then, Algorithm 3 computes the (unique) optimal schedule. The complexity of this algorithm is $\log_2(N)$ times the complexity of solving the one-dimensional equation (7.32).*

Let us comment on the computational complexity of Algorithm 3. First, it is clear that it halts after no more than $\log_2 N$ iterations. Then, for each

Algorithm 3: Dichotomy over the set of speeds for the computation of the optimal schedule.

Input: The set of speeds $s_1, \dots, s_N$, the power function Φ, the deadline D, and the probability distribution of the task size F.
Output: The optimal subset of speeds $s_U, \dots, s_N$, and schedule $y_U, \dots y_N$.

$U := N; L := 1;$
while $U > L$ **do**
  $m := \lfloor (U + L)/2 \rfloor;$
  Solve (7.32) for y_m using speeds $s_m, \dots, s_N$, where $y_{m+1}, \dots, y_N$ are given by (7.30);
  if $y_m \leq 0$ **then**
    $L := m;$
  else
    $U := m;$
  end
end

iteration, a one-dimensional equation needs to be solved, i.e., Eq. (7.32). Rearranging terms and using Eq. (7.30), this equation can be rewritten as

$$D = \frac{W_{\max}}{s_N} + \sum_{i=m}^{N-1} y_i S_i = \frac{W_{\max}}{s_N} + \sum_{i=m}^{N-1} (F^c)^{-1} \left(\frac{\Gamma_m}{\Gamma_i} F^c(y_m) \right) S_i, \tag{7.33}$$

where $S_i \stackrel{\text{def}}{=} \frac{1}{s_i} - \frac{1}{s_{i+1}}$.

We notice that Equation (7.33) admits a unique solution, and that F is differentiable almost everywhere because it is increasing. If derivatives can be computed, one can solve Equation (7.33) by using (efficient) standard root-finding algorithms such as the Newton–Raphson or the secant method.

7.5 Multiple Stochastic Jobs

This section shows how to compute the optimal scheduler for *multiple jobs* with random features. A similar question has been addressed in full details in [11]. Here, however, we consider a more general case. More precisely, EDF is chosen *a priori* to schedule jobs in [11], while we consider the general case. We prove later in the section that EDF is indeed an optimal choice for scheduling jobs.

In this section, all the time quantities (arrival times, execution times, deadlines) are random variables with values in $\mathbb{N}$. This assumption can be taxed as too strong. Actually, it is a natural discretization based on the *characteristic time* of the stream of jobs, denoted ξ. This is the greatest duration such that the jobs' features (deadlines and arrival times) are multiple of ξ. This discretization is classical in real-time systems and reasonable. It will play an essential role in our design of a discrete-time scheduler. The time unit ξ only depends on the jobs' features, and is unrelated to the cycle time of the processor, which can be several orders of magnitude smaller. In the following, we assume that the size and the deadline of the next job does not depend on time and so have *stationary distributions*. The triplets (τ_i, w_i, d_i) of all jobs are random variables, defined on a common probability space. We assume that the joint distribution is known, i.e., $\forall t, w, d$, $\mathbb{P}(\tau_i = t, w_i = w, d_i = d)$ is supposed to be known. We assume also that the distribution of inter-arrival times has a finite memory bounded by L. We note ℓ_t the time elapsed between the last job arrival and t. Finally, we focus on the *clairvoyant* case, in contrast with Section 7.4: the size w_i of a job and its deadline d_i are known to the scheduler at the release time r_i of the job J_i.

The role of the scheduler (or controller) is to choose which job to execute among all currently pending jobs and the current speed for the processor. The scheduler acts at discrete times compatible with the time unit ξ used to quantify the job features, i.e., it acts at every time $t\,\xi$, with $t \in \mathbb{N}$. In the sequel, we will forget about ξ and simply take as the time scale the set of integers $\mathbb{N}$. We further assume that jobs are *preemptive* with a null switching time. Our construction of the scheduler is given in Section 7.5.2. It can be modified to relax this null switching time assumption, as we will see in Section 7.5.3.1.

Here again, we focus on the case with a discrete number of speeds: the processor can run at any time t at a speed $s(t)$ belonging to a finite set of integer speeds $\mathcal{S}$ whose largest element is $s_{\max}$. The power consumption Φ is *arbitrary* (convexity is not required). However, several structural properties (such as monotonicity) will only hold when the function Φ is convex. Finally, Subsection 7.5.3.1 briefly explains how it is possible to take into account time lags for speed changes, and preemption times to generalize our model.

7.5.1 Problem statement

Here, the objective function is the long-run energy consumption while satisfying job deadlines for an infinite stream $\mathcal{J} = (J_i)_{i \in \mathbb{N}}$ of jobs. A precise definition of the objective function is given in Equation (7.37). This objective is challenging because it mixes a random environment with hard deterministic constraints. One solution is to replace the hard deterministic constraints by soft random constraints: deadlines become random, and hard constraints are placed by missed deadline penalties. This has been done in [3]. In that case, the optimal policy can be computed in also closed form. Here, we consider the more difficult problem where deadlines have to be met with absolute certainty, inducing the question of feasibility.

As mentioned before, the information available to the processor only changes at integer point, since all features are integers. We may assume that the decision times, i.e., instants when the processor can change its speed and executed jobs, are integers. This can be done with no loss of optimality as shown in [10]: Letting the scheduler change the speed and executed jobs at any point in time will not help to reduce the energy consumption w.r.t. speed and job changes only occurring at integer times.

Our problem is therefore a constrained optimization problem (minimize the energy while satisfying all the deadlines), which looks hard to solve at a first glance. Actually, it can be modeled as an unconstrained Markov Decision Process (MDP) by introducing a specific state space that also encodes the constraints of the problem. The proposed solution is explained in the next subsection. It achieves optimal energy performance (on average) as well as optimal feasibility.

7.5.2 Markov Decision Process solution

In this section, we design a Markov decision process (MDP) model of our problem. We construct its state space $\mathcal{X} \times \{0, \ldots, L\}$, its action space $\mathcal{C}$ and its transition probabilities P.

7.5.2.1 Backlog staircase function used as state

As in Section 7.3.2, the information needed to compute the optimal decision at time t is the pending work or *backlog work* (cf. Def. 7.11). This will serve as the state of the MDP. We recall its definition here:

DEFINITION 7.11. The backlog work at time t is an integer valued càdlàg[3] staircase increasing function $\beta_t(\cdot)$, where $\beta_t(u)$ is the amount of work arrived before t that must be completed before time $t + u$.

With this representation, we can notice that the choice of the next decision at time t may only depend on the backlog work function β_t associated with the elapsed time since the latest arrival, $\ell_t \in \mathbb{N}$. In the following, we call $\boldsymbol{x}_t = (\beta_t, \ell_t)$ the *state* of the system at time t.

The state space is finite: both β and ℓ are bounded. Its size is the number of staircase functions with support $\{0, \ldots, d_{\max}\}$ whose steps are bounded by $W_{\max}$. A combinatorial argument exposed in [11] shows that it grows roughly as $(W_{\max})^{d_{\max}}$.

7.5.2.2 Actions and state space transitions

Recall that we have defined in Section 7.3.1.3 the basic operators $\mathbb{T}f$ (time shift), f^+ (positive part), and H_t (Heaviside function) on the one hand in Def. 7.3, and the set $\mathcal{A}_t$ (jobs released at t) and the quantity $A_t(u)$ (amount of work arrived at t that must be executed before $t + u$) on the other hand in Def. 7.4.

In a dual fashion, the set $\mathcal{C}(s)$ is the set of pending jobs that the scheduler

[3] Càdlàg = continue à droite, limite à gauche = right continuous with left limits.

chooses to partially execute (c_i units of job i is executed) in the next time interval:

$$\mathcal{C}(s) := \{(c_i, \delta_i), J_i \text{ with relative deadline at } \delta_i\}. \tag{7.34}$$

The speed of the processor is the total work executed over one time interval, namely $\sum_i c_i$. The next definition formalizes the choice of the scheduler at any time t. The service execution from $t-1$ to t and the corresponding speed:

$$C(u) := \sum_{(c_i,\tau_i)\in\mathcal{C}(s)} c_i H_{\delta_i}(u), \quad s(C) := C(d_{\max}).$$

Lemma 7.1 introduces the transition formula from one state to another.

LEMMA 7.1. *Under this framework, the backlog at time t is a Markovian function of the backlog at time $t-1$:*

$$\begin{aligned}\beta_t(\cdot) &:= \mathit{Succ}(\beta_{t-1}(\cdot), C_{t-1}, A_t)\\ &:= \mathbb{T}\left[(\beta_{t-1}(\cdot) - \mathbb{T}^{1-t}C_{t-1}(\cdot))^+\right] + A_t(\cdot).\end{aligned} \tag{7.35}$$

The backlog function β does not contain any information about the time elapsed since the latest arrival. This information, denoted ℓ_t, must be added to get the full state of the system at time t. The relationship between ℓ_t and ℓ_{t-1} is as follow:

$$\ell_t = \mathit{Succ}(\ell_{t-1}, A_t) := \begin{cases} 0 & \text{if } A_t \neq 0\\ (\ell_{t-1}+1)\wedge L & \text{otherwise.}\end{cases} \tag{7.36}$$

To compute the optimal policy π^*, we take advantage of the *Markovian evolution* of the state. From the distribution of jobs (τ_i, w_i, d_i), one can build, under state $\boldsymbol{x}$ and at time t, the probability distribution of the work that arrives at t. For any $(w_1, \ldots, w_{d_{\max}})$:

$$\boldsymbol{p}_{\boldsymbol{x}}(w_1, \ldots, w_{d_{\max}}) := \mathbb{P}\left(A_t = \sum_{k=1}^{d_{\max}} w_k H_{k+t} \mid \boldsymbol{x}\right).$$

Once $\boldsymbol{p}$ is given, the transition matrix $P(\boldsymbol{x}, C, \boldsymbol{x}')$ from state $\boldsymbol{x} = (\beta, \ell)$ to $\boldsymbol{x}' = (\beta', \ell')$ when service chosen by the processor is C is:

$$P_t(\boldsymbol{x}, C, \boldsymbol{x}') = \begin{cases} \boldsymbol{p}_{\boldsymbol{x}}(w_1, \ldots, w_{d_{\max}}) \\ \qquad \text{if } \beta' = \mathit{Succ}(\beta; C, A_t) \\ \qquad\qquad \text{and } \ell' = \begin{cases} 0 & \text{if } A_t \neq 0 \\ (\ell+1)\wedge L & \text{otherwise} \end{cases} \\ 0 \quad \text{otherwise.} \end{cases}$$

This shows that the transition probability can be expressed as a function of the probability distributions of the jobs, through the distribution $\boldsymbol{p}$. If jobs are independent, then $\boldsymbol{p}$ can be computed using a convolution of the job distributions.

Of course, the total energy consumption over an infinite time horizon is infinite whatever the speed policy is. This is why we minimize the *long-term average* energy consumption *per time-unit*, denoted v. We therefore look for the optimal policy π^* that minimizes the long-run energy consumption as defined in Equation (7.37):

$$v^* = \min_{\pi} \mathbb{E}\left(\lim_{T\to\infty} \frac{1}{T}\sum_{t=1}^{T} \Phi(\pi(\boldsymbol{x}_t))\right). \tag{7.37}$$

7.5.2.3 Value Iteration Algorithm

Unichain Assumptions (A_1)**:** We assume that the probability that no job arrives in the next time slot is strictly positive. Starting from an arbitrary system state $\boldsymbol{x}$, with positive probability, no more jobs will arrive for long enough a time so that all past deadlines have been met and the state reaches the empty state $\beta_0 = ((0,\ldots,0), \ell = L)$. This implies that the empty state is recurrent under any policy so that the MDP is *unichain* [21].

From [21], if the MDP is unichain, then the limit in Equation (7.37) always exists; it does not depend on the initial state and it can be computed with an arbitrary precision using the *value iteration* of the Bellman equation, called (VI) here.

The set $\mathcal{F}(\boldsymbol{x})$, defined in Equation (7.38), is the set of *feasible choices* in state $\boldsymbol{x}$. The feasible choices are those jobs and speed selection that are good enough to ensure to execute the urgent backlog work at the next time step. For state $\boldsymbol{x} = (\beta, \ell)$,

$$\mathcal{F}(\boldsymbol{x}) = \{C \text{ s.t. } C \geq \beta(1)H_1 \text{ and } s(C) \in \mathcal{S}\}. \tag{7.38}$$

In the condition of the while loop of Algorithm 4, the *span* of a vector z is the difference between its maximal value and its minimal value: $span(z) = \max_i(z_i) - \min_i(z_i)$. A vector with a span equal to 0 has all its coordinates equal.

In Algorithm 4, the quantity u^n can be seen as the total energy up to iteration n. Algorithm 4 computes both the optimal average energy consumption per time unit (g^*) with a precision ε as well as an ε-optimal speed to be selected in each state $(\pi^*[\boldsymbol{x}])$.

The time complexity to compute the optimal policy depends exponentially on the precision $\frac{1}{\varepsilon}$.

At runtime, the processor computes the current state $\boldsymbol{x}_t$ and performs a Table Look-Up (TLU) algorithm to obtain its optimal speed $\pi^*[\boldsymbol{x}_t]$, the speed tables having been computed offline by (VI). This algorithm, called TLU_{VI}, is shown in Algorithm 5. The size of the table is $Q(d_{\max})$.

Algorithm 4: Value Iteration Algorithm (MP) to compute the optimal speeds in each state and the average energy cost per time unit.

$u^0 \leftarrow (0, 0, \dots, 0)$, $u^1 \leftarrow (1, 0, \dots, 0)$
$n \leftarrow 1$
$\varepsilon > 0$ % sub-optimality error
while $span(u^n - u^{n-1}) \geq \varepsilon$ **do**
  for all $\boldsymbol{x} \in \mathcal{B}$ **do**
    $u^{n+1}(\boldsymbol{x}) \leftarrow \min_{C \in \mathcal{F}(\boldsymbol{x})} \left(\Phi(s(C)) + \sum_{\boldsymbol{x}' \in \mathcal{B}} P(\boldsymbol{x}, C, \boldsymbol{x}') u^n(\boldsymbol{x}') \right)$
  end for
  $n \leftarrow n + 1$
end while
Choose any $\boldsymbol{x} \in \mathcal{B}$ and let $g^* \leftarrow u^n(\boldsymbol{x}) - u^{n-1}(\boldsymbol{x})$
for all $\boldsymbol{x} \in \mathcal{B}$ **do**
  $\pi^*[\boldsymbol{x}] \leftarrow \arg\min_{C \in \mathcal{F}(\boldsymbol{x})} \left(\Phi(s(C)) + \sum_{\boldsymbol{x}' \in \mathcal{B}} P(\boldsymbol{x}, C, \boldsymbol{x}') u^n(\boldsymbol{x}') \right)$
end for
return π^*

Algorithm 5: Runtime process: Table Look-UP, TLU_{VI}, used by the processor to apply the optimal speed.

For each $t = 0 \dots T - 1$
  Update $\boldsymbol{x}_t$ using Equation (7.35)
  Set $C := \pi_t^*[\boldsymbol{x}_t]$
  Execute the job(s) according to
    the choice function C for one time unit
End

The schedulability condition (A_2): A set of jobs $\mathcal{J}$ is *schedulable* if there exists a scheduling policy that completes all jobs before their deadline.

This induces the notion of *schedulability under a policy* π: If using speed $\pi(\cdot)$, the processor executes all jobs before their deadlines, then $\mathcal{J}$ is schedulable under π.

THEOREM 7.6. *Under assumptions (A_1) and (A_2), the policy π^* computed by Algorithm 4 and used as described in Algorithm 5 never misses a deadline and is ε-optimal for energy minimization.*

Proof. (**Sketch**) Energy optimality is a direct consequence of MDP theory: Algorithm 4 gives the energy spent per time-unit in the long run, v^* up to the arbitrary additive precision ε.

As for the schedulability under π^*, the construction of the feasibility set $\mathcal{F}(\boldsymbol{x})$ makes sure that, if the state reached at time t is $\boldsymbol{x}$, then no job will miss its deadline in the next time step, unless the set $\mathcal{F}(\boldsymbol{x})$ is empty. Repeating this argument at each time step gives the proof as soon as one shows that the set $\mathcal{F}(\boldsymbol{x})$ never becomes empty under the optimal policy. By definition, the set $\mathcal{F}(\boldsymbol{x})$ can only be empty when $s_{\max}$ is not large enough.

Let us assume that the set of jobs is schedulable. This means that there exists a policy that can schedule all jobs before their deadline using speeds in $\mathcal{S}$. Let us remove the constraint $C(d_{\max}) \leq s_{\max}$ in $\mathcal{F}(\boldsymbol{x})$ and let us extend the definition of the power function: $\Phi(s) = \infty$ if $s > s_{\max}$. Now, under any state $\boldsymbol{x}$, the new set $\mathcal{F}(\boldsymbol{x})$ is not empty under the optimal policy. The existence of a policy never missing deadlines says that under the new framework, this policy has a finite cost (same cost as under the original system). This implies that the cost of the optimal policy is also finite, hence it does not use speeds larger than $s_{\max}$. This implies schedulability under the optimal policy. □

7.5.3 Additional properties and generalizations

The first property is the optimality of EDF as far as job selection is concerned. The action of the scheduler $C(s)$ is which jobs are being served and the total amount of service (s). The next results say that there is an optimal policy that always serves jobs with the smallest relative deadline.

PROPOSITION 7.5. *There exists an optimal policy whose job selection is done using EDF.*

Proof. (**Sketch**) Let π^* be an optimal policy that does not always schedule jobs with EDF. Let us call $\boldsymbol{x}_0$ a state under which π^* is not EDF.

Let us construct a new policy π' as follows. In state $\boldsymbol{x}_0$, the execution choice is $C'(\boldsymbol{x}_0)$ that executes the same amount of work as C^* but uses EDF to select the jobs being executed instead. Note that $C'(\boldsymbol{x}_0)$ is feasible, being larger than $C^*(\boldsymbol{x}_0)$.

With the new policy, the next state under arrival A is $\boldsymbol{x}'_1 := Succ(\boldsymbol{x}, \pi'(\boldsymbol{x}), A)$ and is smaller than $\boldsymbol{x}^*_1 = Succ(\boldsymbol{x}, \pi(\boldsymbol{x}), A)$, for any A: $\boldsymbol{x}'_1(t) \leq \boldsymbol{x}^*_1(t)$, for all t. Since $\boldsymbol{x}'_1 \leq \boldsymbol{x}^*_1$, then the feasibility sets under both states are also comparable: $\mathcal{F}(x^*_1) \subseteq \mathcal{F}(x'_1)$.

In state $\boldsymbol{x}^*_1$, the optimal policy uses action C^*_1. One can construct an action C'_1 used for state $\boldsymbol{x}'_1$ that mimics C^*_1 as much as possible in the following sense: The execution function C'_1 is the largest function smaller than C^*_1 and smaller than $\boldsymbol{x}'_1$. This choice has several properties:

1. The executed work of both policies is the same: $s(C'_1) = s(C^*_1)$, hence both have the same power consumption.

2. If C^*_1 is EDF, then C'_1 is also EDF.

3. For any A (the same A is used under both policies),

$$\boldsymbol{x}'_2 := Succ(\boldsymbol{x}'_1, C'_1, A) \leq \boldsymbol{x}^*_2 := Succ(\boldsymbol{x}^*_1, C^*_1, A).$$

4. The new jobs (i.e., jobs in A) are executed identically under C_1^* and under C_1'.

This construction can be carried for $d_{\max}$ steps, yielding comparable states $\boldsymbol{x}_i' \leq \boldsymbol{x}_i^*$, $i = 0...d_{\max}$, and one additional property after $d_{\max}$ steps: All initial jobs forming the state $\boldsymbol{x}$ have been executed under both policies, and the jobs that arrived during these steps have been executed identically under both policies. This implies that $\boldsymbol{x}'_{d_{\max}} = \boldsymbol{x}^*_{d_{\max}}$. After this point, the new policy is the same as the optimal policy π^*. Since the energy consumption is the same under both policies, then π' is also optimal and uses EDF in at least one more state than π^*. Eventually, this implies the existence of a non-stationary optimal policy that uses EDF.

Now, among the policies using EDF to execute the jobs, the best policy is stationary by dominance of stationary policies in finite MDPs (see [21]).

Combining this with the previous result shows the existence of an optimal stationary policy that uses EDF. □

Proposition 7.5 has several practical implications: One can limit the choice of the scheduler to the choice of the speed, s, the choice of served jobs being implicitly made by EDF, and the feasibility condition is replaced by a simpler condition: $\mathcal{F}(\boldsymbol{x}) = \{s \in \mathcal{S}, s \geq \beta(1)\}$. The reduction of the action space also makes computations in (VI) easier and faster.

Next, we show that one compares the (MDP) solution (MP) with the optimal available policy (OA) from [26] introduced in Section 7.3.2. We recall that (OA) chooses the smallest speed such that all the jobs already present in the current backlog work could be executed without a new job arrival. Proposition 7.6 gives the property between (MP) and (OA) speed at a given state.

PROPOSITION 7.6 ([11]). *If Φ is a convex function, for a given state $\boldsymbol{x}$, the optimal speed policy π^* (the speed obtained by* (MP) *policy) is always larger or equal to the speed of* (OA).

Thus, for a given state, the speed of (OA) policy is always larger or equal to the (MP) policy, meaning that the (MP) policy will anticipate the future work arrival by executing more work quantity (with a higher speed choice) at the beginning. However, in the future, it will be in a state that leads to less energy consumption. In fact, (MP) leverages statistical information to temporarily expend more energy at a given moment, allowing it to reduce energy consumption in the future when a potentially higher workload may arise.

Another property we have on (MP) is the monotonicity property. The focus here is to compare two sets of jobs, and search energy consumption relation between these two sets of jobs when we apply the (MP) algorithm. To compare the sets of jobs $\mathcal{J}_1$ and $\mathcal{J}_2$, since we do not know the exact value of their features, we can only compare their distributions (F_1 and F_2 for both jobs independent of the current state $\boldsymbol{x}$).

We define a stochastic order (denoted $\leq_s$) between two sets of jobs $\mathcal{J}_1$ and $\mathcal{J}_2$ (with respective distributions F_1 and F_2) as follows: $\mathcal{J}_2 \leq_s \mathcal{J}_1$ if we can compare the two distributions. In detail, $\forall J_1(\tau_1, w_1, d_1)$ with distribution F_1, $\forall J_2(\tau_2, w_2, d_2)$ with distribution F_2, the probability of arrival is the same, and the distribution of the size and the deadline satisfy the following order:

$$\begin{aligned} &\forall \gamma, \delta, \mathbb{P}(w_2 \geq \gamma, d_2 \leq \delta) \leq \mathbb{P}(w_1 \geq \gamma, d_1 \leq \delta) \\ &\qquad \text{and } \forall t \in \mathbb{N}, \mathbb{P}(\tau_1 = t) = \mathbb{P}(\tau_2 = t). \end{aligned} \tag{7.39}$$

Moreover, by denoting $(i_1^1, \ldots, i_{d_{\max}}^1)$ the work quantity that arrives at time t for $\mathcal{J}_1$, and $(i_1^2, \ldots, i_{d_{\max}}^2)$ the work quantity that arrives at time t for $\mathcal{J}_2$, we have:

$$\begin{aligned} &\forall t, i_1^1, \ldots i_{d_{\max}}^1, i_1^2, \ldots i_{d_{\max}}^2, \\ &\qquad \mathbb{P}(t, \beta_{t+1} \geq i_1^1, \ldots, \beta_{t+d_{\max}} \geq i_{d_{\max}}^{d_{\max}}) \\ &\qquad \leq \mathbb{P}(t, \beta_{t+1} \geq i_1^2, \ldots, \beta_{t+d_{\max}} \geq i_{d_{\max}}^2). \end{aligned}$$

Proposition 7.7 gives the monotonicity properties of the (MP) solution.

PROPOSITION 7.7 ([11]). *If $\mathcal{J}_2 \leq_s \mathcal{J}_1$, then the corresponding long-run energy consumption under π^* satisfies $v^{(2)} \leq v^{(1)}$.*

7.5.3.1 Speed changes and context switch

To make this model more realistic, one can introduce several generalizations that can be included inside our model and that are detailed in [11]. The first one is the cost of speed changes for the processor. So far, it was assumed to be null. However, one can incorporate arbitrary costs for switching from one speed to another in our model. The idea is to transform a time delay due to the speed change into an energy additional cost. The second possible generalization is to take into account the cost endured to perform a context switch between one task to another. The idea is the same; one needs to compensate the delay introduced by the context switch into an additional energy cost in the function Φ.

7.5.3.2 Non-clairvoyant case

The multi-job case has also been explored for the non-clairvoyant case in [14]. In this case, only the worst-case size of the job with size probability distribution is known when the job arrives, but the actual size is not known to the scheduler until the job terminates. This case is more complex and will not be studied in this chapter; for more information, refer to [14]. Also note that in the non-clairvoyant case, EDF is no longer optimal for energy minimization.

7.6 Reinforcement Learning Approach

In this section, we present a machine learning approach to compute a near optimal scheduler when the features (inter-arrival, size and deadline distributions), used in the previous section to compute expectations, are unknown to the scheduler. Instead, job features are *observed* over time, and the speed of the processor is adjusted over time based on these collected past observations. Eventually, an *estimated* Markov Decision Process (MDP) is constructed and a speed policy is computed that would be optimal should the future jobs conform to the estimated statistics. Then, this speed policy is used for subsequent jobs. Of course, the quality of this statistical model depends on the number of observations; therefore, finding a stopping criterion for the learning phase is important as it affects the overall performance, e.g., the *regret* as defined below.

Similar to what happens when the statistics are known to the scheduler, when hard constraints are replaced by soft constraints, the learning problem becomes easier. It has been investigated in [2]. The authors show that the regret (defined in Equation (7.51)) of their learning algorithm is independent of the size of the state space. This section is rather dedicated to the hard deadline case, more aligned with the previous sections.

7.6.1 Learning strategy

In the following, we consider the case where no *a priori* information is available on the job features. However, we assume that the arriving process is iid, and job features (inter-arrival times, sizes and deadlines) are discrete and follow some distribution, denoted $\mathbf{q}$, so that a finite MDP can represent the system evolution once the state space is learned. Therefore, the goal is to learn all the parameters of this MDP (state space and transition probabilities). This MDP has already been described in Section 7.5 and will be reformulated here to highlight its non-explorative nature (see more on this later). Our learning approach is model-based. Model-free approaches such as Q-learning are not as efficient here, as it will also be explained later.

To estimate the MDP parameters, we start with a training part (also called the "learning phase" in the following) during which the learning algorithm learns the state space and the transition probabilities. As jobs arrive in the system, it discovers the reachable states and estimates the distribution of job features. At some time T, this learning phase ends (a stopping criterion to determine T is given in Section 7.6.5). We then compute the optimal speed policy π^T for the estimated distribution of job arrivals $\mathbf{q}_T$ by using an adapted version of the Value Iteration Algorithm. Finally, in the exploitation phase, the policy π^T is used to select the speed used by the processor when executing

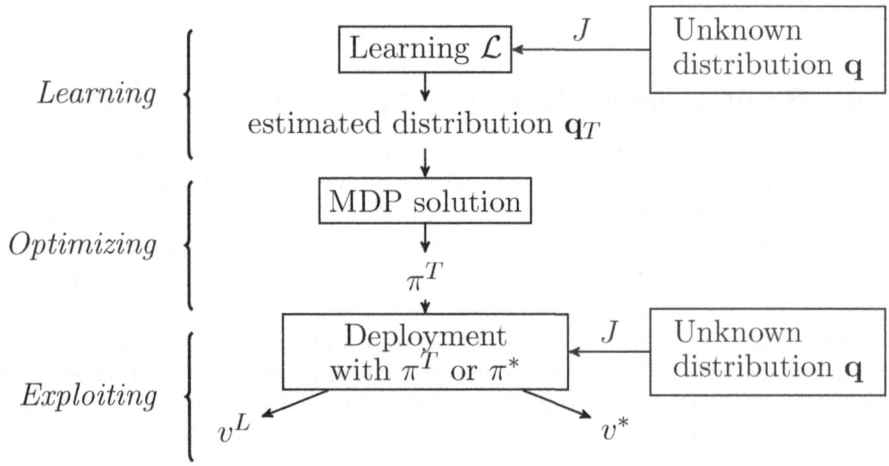

Figure 7.6: Learning procedure to execute jobs while minimizing energy. The procedure is done in three phases: an exploration phase of length T that allows estimation of the transition probabilities, then an optimization phase where we compute a near optimal speed policy π^T (denoted π^T to emphasize the duration of the learning phase), and an exploitation phase that applies this policy on this set of jobs. The performance of this procedure depends on the length of the learning phase and on the performance of the learned policy during the exploitation phase. In the second part of this section, we compare three different policies: the optimal policy π^*, that can only be computed when the probability distribution of jobs is known, the optimal available policy and the policy π^T, based on the learning phase.

jobs arriving according to the true distribution $\mathbf{q}$. This induces an average energy cost denoted v^L. This procedure is illustrated in Figure 7.6.

In most cases (for example in multi-armed bandits), the most efficient learning procedures mix exploration and exploitation, as typically done with Q-learning methods [23] or methods based on upper confidence bounds [4]. Here, however, exploration comes for free: the actions taken by the learner do not alter the sampling process of the job features. More precisely, changing the speed of the processor will not affect the next random sample (arrival time, size and deadline of the upcoming jobs). Therefore, we start with a pure learning phase followed by a pure exploitation phase. We claim this is efficient in our case because of this special type of Markovian system involved, sometimes called non-explorative MDP (see [6]). This fact will be made explicit in the following section, where the vector $\mathbf{q}(a)$ is introduced.

7.6.2 Reformulation of the Bellman equation

In the following, we focus on two arbitrary generic states $\boldsymbol{x} = (\beta, \ell), \boldsymbol{x}' = (\beta', \ell')$, and describe the transition from state $\boldsymbol{x}$ to state $\boldsymbol{x}'$. Since EDF is dominant, we assume in the rest of this section that the policy always uses EDF to schedule jobs and only controls the speed of the server. This simplifies the description of the evolution of the system. We will consider $\mathcal{S}_{\boldsymbol{x}} = \{s \in \mathcal{S}, s \geq \beta(1)\}$, the set of speeds that satisfy the feasibility condition under state $\boldsymbol{x}$.

We recall further notations used for the construction of the simplified MDP:

The *state* of the system at time t is as before the staircase backlog function.

When jobs arrive, the job arrival function at time t, denoted A_t, is defined as before:

$$A_t(\cdot) = \sum_{i \,|\, r_i = t} w_i H_{d_i}(\cdot). \tag{7.40}$$

The set of all possible arrival functions at any time t is denoted $\mathcal{A}$ and is also a staircase function.

In the following, we denote by Succ the simplified successor function: $\boldsymbol{x}' = \text{Succ}(\boldsymbol{x}, s, a)$ denotes the state $\boldsymbol{x}'$ succeeding to state $\boldsymbol{x}$ when the speed s is used over the time interval $[t, t+1]$ and the job arrival function at $t+1$ is a. The formula for $\text{Succ}(\boldsymbol{x}, s, A)$ is

$$\text{Succ}(\beta, s, A) = \mathbb{T}\left[(\beta - s)^+\right] + A, \tag{7.41}$$

where $\mathbb{T}$ is the time shift operator: $\mathbb{T}f(t) = f(t+1)$.

When arrivals follow a probability distribution ($\mathbf{q}(A)$ denotes the probability that the next job arrival is a), this induces a *transition probability*: $\mathbf{P}(\boldsymbol{x}, s, \boldsymbol{x}')$ from state w to state $\boldsymbol{x}'$ under speed s:

$$\mathbf{P}(\boldsymbol{x}, s, \boldsymbol{x}') = \sum_{A : \boldsymbol{x}' = \text{Succ}(\boldsymbol{x}, s, A)} \mathbf{q}(A). \tag{7.42}$$

Since $\text{Succ}(\boldsymbol{x}, s, A)$ is injective in A, the sum has at most one term so one can also write:

$$\mathbf{P}(\boldsymbol{x}, s, \boldsymbol{x}') = \mathbf{q}(A) \text{ if } A \text{ is the unique arrival s.t. } \boldsymbol{x}' = \text{Succ}(\boldsymbol{x}, s, A). \tag{7.43}$$

Under EDF, the *policy* is a simply deterministic speed function $\pi(\boldsymbol{x})$ that selects one speed in each state $\boldsymbol{x}$.

As in Section 7.5, the objective function is the energy spent per time unit:

$$v^{\pi} = \lim_{T \to \infty} \left(\frac{1}{T} \sum_{t=0}^{T} \Phi(\pi(\boldsymbol{x}_t)) \right). \tag{7.44}$$

The value v^* is the unique finite solution (together of a bias vector h^*) of

$$\begin{aligned} v^* + h^*(\boldsymbol{x}) &= \min_{s \in \mathcal{S}_{\boldsymbol{x}}} \left\{ \Phi(s) + \sum_{\boldsymbol{x}' \in \mathcal{W}} \mathbf{P}(\boldsymbol{x}, s, \boldsymbol{x}') h^*(\boldsymbol{x}') \right\} \\ &= \min_{s \in \mathcal{S}_{\boldsymbol{x}}} \left\{ \Phi(s) + \sum_{A \in \mathcal{A}} \mathbf{q}(A) h^*(\mathrm{Succ}(\boldsymbol{x}, s, A)) \right\}. \end{aligned}$$

Note that the version where $\mathbf{q}(A)$ is used in the Bellman equation shows that the random innovation of the system is independent of the current state and current speed.

7.6.3 Learning procedure in a nutshell

Jobs come from an exogenous source, and their features follow an *unknown* distribution. We use observations of the incoming jobs to construct an estimated MDP. The solution of this estimated MDP should be close to the solution of the true MDP with enough observations. Then, this estimated model is used to compute a speed policy to be used to execute new jobs from the same source.

Since the estimation is not perfect, this speed policy will not always take the best speed; however, we show in the following that after a long enough observation period, one can guarantee optimality (or near-optimality) of the corresponding speed schedule.

This procedure is illustrated in Figure 7.6.

The construction of the transition probabilities of the MDP is called PL in the figure. This is a training phase during which we learn the state space and the transition probability matrix. As new jobs arrive, new possible states are reached and the transitions from state to state are updated. At some point when the estimation of the transition matrix is good enough, the learning phase is stopped (a stopping criterion is provided later in the chapter).

A Value Iteration algorithm is used to compute the optimal speed schedules for the learned MDP.

In the application phase, these speeds are used by the processor to execute all incoming jobs.

Let us consider one processor with a variable speed that executes real-time jobs. The goal of this section is to learn the optimal speed policy of the processor, minimizing the energy consumption while completing all jobs before their deadlines under a job sequence that arrives according to some unknown random process.

7.6.4 Learning algorithm $\mathcal{L}$

In this section, we describe the Learning Algorithm $\mathcal{L}$. This algorithm, noted $\mathcal{L}$ from now on, implies that for each couple $(\boldsymbol{x}, s)$, we compute the following equations that depend on one job arrival at time t, A_t:

- Update the state space: $\forall \boldsymbol{x} \in \mathcal{W}, \forall s \in \mathcal{S}_w$, add $Succ(\boldsymbol{x}, s, A_t)$ to $\mathcal{W}$.
- Update the probability vector:

$$\mathbf{q}_{t+1}(A_t) = \frac{t}{t+1}\mathbf{q}_t(A_t) + \frac{1}{t+1} \tag{7.45}$$

$$\mathbf{q}_{t+1}(a) = \frac{t}{t+1}\mathbf{q}_t(a), \text{ if } a \neq A_t \tag{7.46}$$

As defined in the previous section, A_t is the arrival process at time t and Succ$(\boldsymbol{x}, s, A_t)$ is to the next state function w' at time $t+1$. Then, after determining $\mathbf{q}$ inducing the probability transition matrix, we can use the Value Iteration algorithm to compute the optimal policy.

7.6.5 Convergence criterion

In this section, we establish a convergence criterion for the learning based on our estimate on the probability transition matrix. After T steps of algorithm $\mathcal{L}$, we denote by:

- v^* the energy consumption of the optimal policy for the true system (transitions $\mathbf{P}$).
- v^T the energy consumption of the learned policy π^T for the estimated system (transitions $\mathbf{P}_T$).
- v^L the energy consumption of the learned policy π^T for the true system (transitions $\mathbf{P}$).

In the following, we use the L_1 norm for vectors and L_∞ for matrices without further notice.

The stopping criterion is based on two parameters, ϵ and p (typically, both parameters are small). We say that the learning phase stops at time T when, for all random arrival sequences $(A_T)_{t\leq T}$, $\mathbb{P}(|v^* - v^L| > \epsilon) < p$.

A first bound on $v^T - v^*$ can be obtained using perturbation analysis of Markov chains.

LEMMA 7.2. *The error $|v^T - v^*|$ is smaller than ϵ with high probability $1-p$:*

$$\mathbb{P}(|v^T - v^*| < \epsilon) > 1 - p$$

if T, the duration of the learning period, satisfies:

$$T \geq \frac{R_{\max} K(|\mathcal{A}| + \ln(1/p))}{\epsilon^2},$$

where $R_{\max}$ is the energy cost of the maximal speed, $K = \frac{1}{p_0^{d_{\max}}}$ (with p_0 the probability of no job arrival under P_T and $d_{\max}$ the maximal deadline of all the jobs).

Proof. We first bound the difference between the values v^T and v^* by the norm of the difference between two transition matrices of the true system and its estimation. Then, this matrix difference is bounded by using Weissman's inequality [24].

The proof is based on the absolute stability of $\mathbf{P}_T$. By definition, $v^* = \mu^* R$ with μ^* the stationary probabilities under the optimal policies for the exact distribution and $R(w)$ is the power consumption in state w. Using the speed policy of the exact distribution under the estimated transition matrix gives a larger average cost: $v^T \leq \mu^T R^*$, where μ^T is the stationary probabilities of matrix $\mathbf{P}_T(\pi^*)$.

Let $p_0 > 0$ be the estimated probability that no job arrives at the current time ($p_0 = \mathbf{q}_T(\mathbf{0})$). Then, from any state w, the empty state $\mathbf{0} = (0, 0, \ldots, 0)$ is reached after $d_{\max}$ steps with probability larger than $p_0^{d_{\max}}$. This implies that $\mu^T(\boldsymbol{y}0) > p_0^{\mathbf{q}}$. Using the results and terminology presented in [16], the *absolute stability* of $\mathbf{P}_T$ can be written as $\|\mu^* - \mu^T\| \leq K\|\mathbf{P}_T(\pi^*) - \mathbf{P}(\pi^*)\|$ with $K \leq \|B^{-1}\|$. Here, B is the non-singular matrix $\mathbf{I}-(\mathbf{P}_T|\mathbf{0})^{d_{\max}}$, where $\mathbf{P}_T|\mathbf{0}$ is the sub-matrix of $\mathbf{P}_T(\pi^*)$ without its first row and first column. Since for any state $w \neq \mathbf{0}$, $\sum_{w' \neq \mathbf{0}} (\mathbf{P}_T)^{d_{\max}}(w, \pi^*(w), w') < 1 - p_0^{d_{\max}}$, as explained above, then the matrix $\mathbf{U} = \frac{1}{1-p_0^{d_{\max}}} (\mathbf{P}_T|\mathbf{0})^{d_{\max}}$ is a sub-stochastic matrix. The inverse of $\mathbf{B}$ is

$$\mathbf{B}^{-1} = \sum_{n=0}^{\infty} (\mathbf{P}_T|(\mathbf{0})^{d_{\max} n} = \sum_{n=0}^{\infty} (1 - p_0^{d_{\max}})^n \mathbf{U}^n. \tag{7.47}$$

Since $\mathbf{U}$ is sub-stochastic, $\|\mathbf{B}^{-1}\| \leq 1/p_0^{d_{\max}}$.

Finally,

$$\begin{aligned} v^T - v^* &\leq \mu^T R - \mu^* R \\ &\leq \|\mu^* - \mu^T\| R_{\max} \\ &\leq \frac{R_{\max}}{p_0^{d_{\max}}} \|\mathbf{P}_T(\pi^*) - \mathbf{P}(\pi^*)\|. \end{aligned}$$

The opposite term is treated similarly:

$$v^* - v^T \leq \frac{R_{\max}}{p_0^{d_{\max}}} \|\mathbf{P}_T(\pi^T) - \mathbf{P}(\pi^T)\|.$$

The error depends on the norm $\|\mathbf{P}_T(\pi) - \mathbf{P}(\pi)\|$, for $\pi = \pi^*$ or $\pi = \pi^T$.

For any speed s, each entry $\mathbf{P}_T(w, s, w')$ (resp. $\mathbf{P}(w, s, w')$) is equal to $\mathbf{q}_T(a)$ (resp. $\mathbf{q}(a)$) when there exists a such that $\mathrm{Succ}(w, s, a) = w'$ (in which case, it is unique) and 0 otherwise. Therefore, the L_1 norm $\|\mathbf{P}_T(\pi) - \mathbf{P}(\pi)\|$ for all π is equal to the L_1 norm of the vector $\mathbf{q}_T - \mathbf{q}$.

By definition, $\mathbf{q}_T(a)$ is the empirical mean of t independent random variables in $\{0, 1\}$. The Weissman inequality says, for all $\epsilon > 0$, the L1 norm satisfies

$$\mathbb{P}(\|\mathbf{q}_T - \mathbf{q}\| \geq \epsilon) \leq 2^{|\mathcal{A}|} e^{-T\epsilon^2/2}. \tag{7.48}$$

Therefore, by choosing a small probability p and setting $\exp(-T\epsilon^2/2 + |\mathcal{A}|\log(2)) \leq p$, we get $T \geq \frac{\log(\frac{1}{p})+|\mathcal{A}|}{\epsilon^2}$ since $\log(2) < 1$. □

We are now ready to assess the performance of the learned policy π^T in the true system.

PROPOSITION 7.8. *The error $|v^L - v^*|$ is smaller than ϵ with high probability $1-p$, if T, the duration of the learning period, satisfies:*

$$T \geq \frac{R_{\max} K(|\mathcal{A}| + \ln(1/p))}{\epsilon^2}. \tag{7.49}$$

Proof. It is not easy to compare v^L and v^* directly. Instead, we use

$$|v^L - v^*| \leq |v^L - v^T| + |v^T - v^*|, \tag{7.50}$$

and bound the second term using the previous lemma. As for the first term, the proof is similar to the proof of the previous lemma (a little easier because a single policy is used).

Let μ^L be the stationary measure of $\mathbf{P}(\pi^T)$ and μ^T be the stationary measure of $\mathbf{P}_T(\pi^T)$. By definition of the average cost, $v^L = \mu^L R^T$ and $v^T = \mu^T R^T$.

Therefore,

$$\begin{aligned} |v^L - v^*| &= |\mu^L R^T - \mu^L R^T| \\ &\leq \|\mu^L - \mu^L\| R_{\max} \\ &\leq \frac{R_{\max}}{p_0^{d_{\max}}} \|\mathbf{P}_T(\pi^T) - \mathbf{P}(\pi^T)\|. \end{aligned}$$

The L_1 norm $\|\mathbf{P}_T(\pi^T) - \mathbf{P}(\pi^T)\|$ has already been bounded in the proof of the previous lemma.

Therefore, the bound for the first term in Equation (7.50) is the same as for the second one. The global bound is twice the bound given in the previous lemma. □

7.6.6 Bounded regret

As discussed in the introduction of this section, the current MDP is non-explorative. This should help to control the regret of the learning algorithm $\mathcal{L}$. Let us first define the *Bellman gap* of a state-action pair (x, s) as

$$B(\boldsymbol{x}, s) := \Phi(s) - \Phi(\pi^*(\boldsymbol{x})) + \mathbf{P}(\boldsymbol{x}, s)h^* - \mathbf{P}(\boldsymbol{x}, \pi^*(\boldsymbol{x}))h^*.$$

This Bellman gap is always non-negative and equal to 0 if and only if the action s is optimal under state $\boldsymbol{x}$. In general, it measures the local sub-optimality of action s in state $\boldsymbol{x}$ if all subsequent actions are optimal.

The *regret* (sometimes called pseudo-regret) of Algorithm $\mathcal{L}$ on MDP M after T steps is classically defined as:

$$\text{Reg}(T, \mathcal{L}, M) := \sum_{t=1}^{T} B(\mathbf{X}_t, s_t), \tag{7.51}$$

where $\mathbf{X}_t$ is the state visited at time t, and s_t the speed chosen by the algorithm.

The main idea behind the following theorem is to exploit Proposition 7.8 to bound the regret of $\mathcal{L}$. A policy is called sub-optimal if $v^\pi < v^*$. Let us now define the sub-optimality gap $G_{\min}$ of the MDP M by:

$$G_{\min} := \min_{\pi \text{ sub-optimal}} (v^* - v^\pi).$$

The sub-optimality gap is strictly positive because the number of policies is finite.

The following theorem is the main result of this section:

THEOREM 7.7. *With high probability* $1 - p$, *the regret of* $\mathcal{L}$ *is bounded:*

$$\text{Reg}(T, \mathcal{L}, M) \leq \frac{R_{\max} K(|\mathcal{A}| + \ln(1/p))}{G_{\min}^2} (R_{\max} + sp(h^*)). \tag{7.52}$$

Proof. Choosing ϵ smaller than the sub-optimality gap in Proposition 7.8 to choose T makes sure that the learned policy π^L is optimal. After this point, all Bellman gaps are null, so the regret is bounded by the regret over the first T steps. At each step, the instant regret $B(X_t, s_t)$ is obviously bounded by $R_{\max} + sp(h^*)$, where $sp(h^*)$ is the span of vector h^*. □

Practical considerations.

The sub-optimality gap $G_{\min}$ may be hard to evaluate. In this case, the bound on the regret given in Theorem 7.7 has little practical interest. If we focus instead on Equation (7.49) for the duration of the learning period with a fixed ϵ, notice that it only uses terms that are known or can be computed during the execution of the learning program: $\mathcal{A}$, p_0, and $d_{\max}$ can be updated at each job arrival, while $R_{\max}$ is a fixed parameter of the model.

The most critical term is $|\mathcal{A}|$, the number of possible arrivals. When no simultaneous arrivals are possible, a rough bound is $|\mathcal{A}| \leq C_{\max}^d$, where C is the maximal size of a job. Under a more realistic case where jobs come from a finite set of M periodic and faulty tasks, $|\mathcal{A}| \leq 2M$ when all tasks have the same period distribution.

To give a practical estimate of the length of the learning period, let us consider the case where one wants to satisfy $v^L - v^* \leq 0.01$ in 99% of the simulations. Then, according to our bound, the training period must use at least $T \geq KR_{max}(|\mathcal{A}| + \log(100))10^4$ job arrivals. In most cases used for numerical experiments, R_{max} is close to 100, and K is close to 10. As for $|\mathcal{A}|$, this is around 10. In total, this amounts to $T \geq 10^8$. This imposes a very long training period to guarantee a standard error in the value. Numerical experiments show that in most cases, a short training period is enough to get very good performance. This may come from two reasons: first, our theoretical bounds are not tight, and second, two quite different transition matrices may have similar optimal costs: under the true as well as the learned system, the lower the speed, the better the energy cost. Therefore, for any given state $\boldsymbol{x}$, both optimal policies π^T and π^* will use the smallest speed most of the time as long as they do not violate the feasibility condition in $\boldsymbol{x}$.

7.7 Conclusion

In this chapter, we have analyzed a fundamental scheduling problem: how a server with variable processing speeds can efficiently schedule jobs with hard deadlines while minimizing energy consumption. This problem arises in various contexts, from real-time systems to high-performance and cloud computing platforms. Meeting deadline constraints is crucial from the end-user's perspective, as violations can significantly impact revenue. Conversely, minimizing energy consumption is essential for the platform manager, who aims to reduce operational costs.

We have explored various scheduling policies and analyzed their feasibility and energy efficiency under different assumptions. More specifically, along the sections, the information available to the server becomes increasingly complex. We have shown that this growing complexity requires different mathematical techniques to analyze the problem effectively.

As a next step, several research directions emerge. For instance, incorporating adaptive learning techniques to dynamically refine scheduling policies based on real-time system observations could further improve performance. Also, the integration of energy-aware scheduling within broader resource management strategies – considering network constraints or storage requirements – remains an open challenge with significant practical implications.

7.8 References

[1] J. Anselmi and B. Gaujal. Energy optimal activation of processors for the execution of a single task with unknown size. In *2022 30th International Symposium on Modeling, Analysis, and Simulation of Computer and Telecommunication Systems (MASCOTS)*, pages 65–72, 2022.

[2] J. Anselmi, B. Gaujal, and L. S. Rebuffi. Reinforcement learning in a birth and death process: breaking the dependence on the state space. In *Proceedings of the 36th International Conference on Neural Information Processing Systems*, NIPS '22, Red Hook, NY, USA, 2022. Curran Associates Inc.

[3] J. Anselmi, B. Gaujal, and L. Sébastien Rebuffi. Optimal Speed Profile of a DVFS Processor under Soft Deadlines. *Performance Evaluation*, 152, Dec. 2021.

[4] P. Auer and R. Ortner. Logarithmic online regret bounds for undiscounted reinforcement learning. In *Advances in Neural Information Processing Systems 19, Proceedings of the Twentieth Annual Conference on Neural Information Processing Systems, Vancouver, British Columbia, Canada, December 4-7, 2006*, pages 49–56, 2006.

[5] N. Bansal, T. Kimbrel, and K. Pruhs. Speed scaling to manage energy and temperature. *J. ACM*, 54(1):1–39, 2007.

[6] V. Boone. *Optimal Regrets in Markov Decision Processes.* PhD thesis, University Grenoble Alpes, 2024.

[7] S. Boyd and L. Vandenberghe. *Convex Optimization.* Cambridge University Press, March 2004.

[8] A. R. Conn, K. Scheinberg, and L. N. Vicente. *Introduction to Derivative-Free Optimization.* SIAM, Philadelphia, PA, USA, 2009.

[9] F. Filippini, J. Anselmi, D. Ardagna, and B. Gaujal. A stochastic approach for scheduling AI training jobs in gpu-based systems. *IEEE Trans. Cloud Comput.*, 12(1):53–69, 2024.

[10] B. Gaujal, A. Girault, and S. Plassart. A Discrete Time Markov Decision Process for Energy Minimization Under Deadline Constraints. Research Report RR-9309, Grenoble Alpes; Inria Grenoble Rhône-Alpes, Université de Grenoble, Dec. 2019.

[11] B. Gaujal, A. Girault, and S. Plassart. Dynamic speed scaling minimizing expected energy consumption for real-time tasks. *J. of Scheduling*, 23(5):555–574, 2020.

[12] B. Gaujal, A. Girault, and S. Plassart. Feasibility of on-line speed policies in real-time systems. *Real Time Syst.*, 56(3):254–292, 2020.

[13] B. Gaujal, A. Girault, and S. Plassart. A pseudo-linear time algorithm for the optimal discrete speed minimizing energy consumption. *Discrete Event Dynamic System*, 31:163–184, Nov. 2021.

[14] B. Gaujal, A. Girault, and S. Plassart. An mdp-based solution for the energy minimization of non-clairvoyant hard real-time systems. *Real-Time Systems*, pages 1–49, 2024.

[15] B. Gaujal, N. Navet, and C. Walsh. Shortest path algorithms for real-time scheduling of FIFO tasks with minimal energy use. *ACM Trans. Embed. Comput. Syst.*, 4(4), 2005.

[16] I. Ipsen and C. Meyer. Uniform Stability Of Markov Chains. *SIAM J. Matrix Anal. Appl.*, 15(4), 1994.

[17] M. Li and F. Yao. An efficient algorithm for computing optimal discrete voltage schedules. *SIAM J. Comput.*, 35:658–671, 2005.

[18] M. Li, F. F. Yao, and H. Yuan. An $O(n^2)$ algorithm for computing optimal continuous voltage schedules. In *Annual Conference on Theory and Applications of Models of Computation, TAMC'17*, volume 10185 of *LNCS*, pages 389–400, Bern, Switzerland, Apr. 2017.

[19] J. R. Lorch and A. J. Smith. Improving dynamic voltage scaling algorithms with PACE. In *Joint International Conference on Measurements and Modeling of Computer Systems, SIGMETRICS'01*, pages 50–61, Cambridge (MA), USA, June 2001. ACM.

[20] J. Mao, C. Cassandras, and Q. Zhao. Optimal dynamic voltage scaling in energy-limited nonpreemptive systems with real-time constraints. *IEEE Trans. Mob. Comput.*, 6(6):678–688, 2007.

[21] M. L. Puterman. *Markov Decision Processes: Discrete Stochastic Dynamic Programming.* Series in Probability and Statistics. Wiley, Hoboken, USA, 1994.

[22] J. Stankovic, M. Spuri, M. D. Natale, and G. Buttazzo. Implications of classical scheduling results for real-time systems. *IEEE Comput.*, pages 16–25, June 1995.

[23] R. S. Sutton and A. G. Barto. *Reinforcement learning: an introduction.* MIT Press, Cambridge, Mass., 1998. http://www.worldcat.org/oclc/37293240.

[24] T. Weissman, E. Ordentlich, G. Seroussi, S. Verdu, and M. J. Weinberger. Inequalities for the L1 Deviation of the Empirical Distribution. Tech. Report HPL-2003-97R1, HP Labs, 2003.

[25] R. Xu, C. Xi, R. Melhem, and D. Mossé. Practical PACE for embedded systems. In *International Conference on Embedded Software, EMSOFT'04*, pages 54–63, Pisa, Italy, Sept. 2004. ACM.

[26] F. Yao, A. Demers, and S. Shenker. A scheduling model for reduced CPU energy. In *IEEE Annual Foundations of Computer Science, FOCS'95*, pages 374–382, Milwaukee, WI, USA, Oct. 1995. IEEE.

Part IV

Social Impact

Chapter 8

Societal Impacts of Scheduling in HPC and Datacenters

Danilo Carastan-Santos, Christophe Cérin, Jean-Marc Nicod, and Denis Trystram

Our objective within this chapter is to clarify the environmental impacts of high-performance computing (HPC) platforms and datacenters. On one side, these domains contribute significantly to carbon emissions. The total weight of both domains is increasing in absolute terms more rapidly than the gains in energy efficiency. Furthermore, the impact is not just on HPC applications or digital cloud services by themselves, but on the entire ecosystem enabled by digital technology today. On the other side, HPC is helping to reduce the overall environmental footprint in other sectors. Thus, we need to consider the trade-off between the benefits of applications and services and their environmental cost. This is a difficult task since digital technologies are most often hidden and then considered as virtual by the majority of people. In addition, this chapter shows how to develop good practices for energy-efficient architectures and softwares. In particular, we present lifecycle analyses, which make it possible to take everything into account in the carbon footprint, including materiality. One of the fundamental points in our study is to look beyond direct effects and to consider rebound effects when designing new services and applications.

8.1 Introduction

8.1.1 Context

It is well established that the Information and Communication Technologies (ICT) sector has a significant impact on the environmental crisis through the huge energy consumption of computing, storage, and interconnection devices [55]. In particular, the electricity estimate of the sector is about 2 %

in 2023 [83], reaching 4.4 % in the US. The prediction of world datacenter (DC) electricity consumption in 2030 is 3 % [54]. High-performance computing (HPC) is part of ICT, and it impacts the environment not just through carbon emissions, but also significant water use and metal consumption for electronic components, of which only a marginal fraction is recyclable. Computing spans from low-powered Edge devices, Cloud to HPC datacenters, ranging from small- to large-scale platforms. It is challenging to estimate the perimeter of the whole HPC domain.

As we will see in more detail in this chapter, HPC presents an opportunity to address climate change by modeling and simulating scenarios. However, HPC platforms also raise environmental concerns, mainly because of the ever-increasing carbon emissions. The impact of ICT on society is not limited to its environmental aspects. Digitalization is profoundly changing social organizations, raising unprecedented ethical questions and increasing inequalities. We focus on environmental impacts in this chapter.

8.1.2 Objectives and content

Our goal within this chapter is to clarify the environmental impacts of the ICT domain in general and HPC/DC domain in particular, both positive and negative impacts, and thus open up discussions on the current and future role of computing and storage platforms regarding the climate crisis. Our ambition is to draw up an overview of the main problems, identify the levers, and discuss the obstacles to overcoming them. In particular, we provide some elements to estimate the global dynamics of the domain. The organization is as follows:

In Section 8.2, we start by reminding of the findings and basic elements for estimating the overall weight of the field of large-scale computing. Section 8.2.1 describes the global picture of the HPC domain from a historical perspective, based on data from the TOP500 that are available for a long time. Section 8.2.2 presents the main characteristics of these systems and the way they are collected. The focus was historically on performance, but it has evolved to take energy consumption into account, followed by a synthesis of analysis on the environmental impacts in Section 8.2.3. Social impacts are discussed in Section 8.2.4.

The motivation of Section 8.3 is to address the need of tools for the two aspects of measurement (Section 8.3.1) and estimation (Section 8.3.2) of applications that provide elements of sustainability in terms of quantitative environmental impacts. Section 8.3.3 develops how to go further by accounting for the whole life-cycle of a computing infrastructure.

Section 8.4 shows how hardware and infrastructure have been involved in making computing platforms virtual to the users' side. Section 8.4.1 lists how the ICT platform can be greener. Examples are given. Next, Section 8.4.2 presents the evolution toward an HPC-as-a-service model, as the most efficient datacenter.

Section 8.5 presents current efforts to manage ICT infrastructures with carbon-free energy, i.e., totally or partially supplied by renewable energy sources. Given that this energy is intermittent by nature and may be lacking when needed, the many issues to be resolved for the transparent management of such platforms are described in Sections 8.5.1 and 8.5.2. Section 8.5.3 proposes a solution to go further than increasing the energy efficiency again and again, this efficiency having intrinsic limits.

Section 8.6 aims to inform and interact with end users about the positive impact of HPC for the climate change crisis. It recalls some key figures and the challenges to address to make HPC a solution. Section 8.6.1 illustrates the potential gain when resource management takes into account other indicators in a multi-objective perspective. For instance, limiting the energy consumption of an HPC platform makes it possible to maintain the computing service even in the event of a power shortage. Section 8.6.2 shows how execution traces are essential for providing relevant predictions of user behavior (job submissions) to make the HPC platform scheduling process more efficient. Section 8.6.3 briefly discusses some scheduling policies to be implemented in such a context, compared with the traditional First Come, First Served (FCFS) backfilling and other complex strategies.

Section 8.7 presents the main ideas that illustrate the fact that HPC in particular and ICT platforms in general have limited energy efficiency potential. While ICT can be part of the solution to the climate change crisis, appropriate use is mandatory to prevent the sector from becoming one of the main contributors of the crisis, particularly when the main players in the field, although aware of this risk, let Artificial Intelligence (AI) get out of control and make the sector more and more energy-intensive.

Section 8.8 opens the discussion about engaging users in terms of computing sufficiency as the most promising action levers for achieving sustainable computing. We start by presenting the logic behind systemic effects in Section 8.8.1. Then, we show higher-order effects in Section 8.8.2.

Finally, the conclusion in Section 8.9 provides a synthesis of the ideas presented in this chapter.

8.2 HPC and Sustainability: Past and Present

It is often difficult to capture the general dynamics of a domain. For the transportation sector, this is possible through a macroscopic analysis of fuel consumption, but digital components are spread everywhere, and they are most often hidden. For HPC, data from most of the largest computing systems have been collected for many years in the TOP500 [86]. The average characteristics offer a first basis for capturing the entire domain. The very top systems show a different dynamic of evolution, giving us an idea of the growth potential of the domain.

8.2.1 TOP500

The TOP500 was introduced in 1993 to rank the most powerful systems in the world using a common set of applications and criteria. From its archive, the TOP500 highlights the evolution and trends in HPC over the years, including varying parameters on the architecture, location, and computing/memory capacity of systems. Over the years, it has evolved to include other dimensions, such as power consumption or architecture heterogeneity. Among other metrics, an HPC system is identified by name, release year, geographic location, and its *performance*.

It is characterized by theoretical peak performance R_{peak} (based on clock rates) and maximal performance (R_{max}) achieved by running the well-known LINPACK linear algebra benchmark [31].

The architecture dimension includes the number of cores, memory capacities, CPU/GPU types, constructors, etc. Around 2008, energy efficiency became a concern in HPC: the monetary cost of running HPC systems exceeded the cost of purchasing and maintaining them [9]. This highlighted the need to improve the energy efficiency of the upcoming HPC systems.

Regarding technology, a study was published in 2015 [86] using the evolution of R_{max} over time to evaluate the increase in performance compared to Moore's law. It describes key break points in the evolution that highlight a slowdown, starting in 2008 and linked to the growth rate of the number of components. This study is restricted to homogeneous systems and was published before the era of GPUs.

8.2.2 The big picture of benchmarking

In this section, we initiate an inquiry into the material aspects of computer systems and the methodologies employed in benchmarking HPC systems, with a focus on sustainability concerns. The objective is to advocate for a more thorough analysis of sustainability, thereby urging a reevaluation of our current practices.

Computer benchmarking [30] is an established practice, particularly in HPC, using the LINPACK package [31] to run linear algebra codes and measure execution time. The TOP500 machines are benchmarked this way. The LINPACK benchmark consists of a collection of linear algebra routines to solve dense systems of linear equations, and for the TOP500, a scalable version is used to optimize each machine's performance. This method highlights peak performance in solving these equations, although it does not represent overall system performance.

Table 8.1 presents data on the peak performance and energy usage of HPC systems, showing a significant decrease in power per mega floating point

Table 8.1: Computing performance of flagship HPC systems and electrical power. While the last column suggests that HPC systems are becoming increasingly efficient over time in terms of electrical power per megaflops, it should be noted that the 29,581 kW of power for the El Capitan machine represents the annual consumption of a French city of around 90,000 inhabitants, such as Versailles.

Performance scale (flops/s)	**Examples**	**Power**	**Power per megaflops/s**
Megaflops/s = 10^6	Intel 8086/8087, Motorola 68000 (end of 1980s)	1.35 W at ambient 25°C	1.35 W
Gigaflops/s = 10^9	CRAY-2 (1985)	150-200 kW	200 W
Teraflops/s = 10^{12}	ASCI Red (1997)	850 kW	0.85 W
Petaflops/s = 10^{15}	IBM Roadrunner (2008)	2,483 kW	0.002483 W
Exaflops/s = 10^{18}	El Capitan (Nov. 2024)	29,581 kW	0.000029581 W

operations per second (megaflops/s) and a marked increase in the energy needed to operate these systems. Figure 8.1 shows the power consumption in kilowatts on a logarithmic scale, highlighting the consumption of machines, represented by a colored triangle, on the TOP500 list. The Bull Sequana machine we point to in this figure is in the middle of the TOP500 ranking and still has a consumption of 917 kW, which is roughly the same as the top 1 1997 ASCI Red machine, and it is mentioned in Table 8.1. Should we be satisfied with this level of power used nearly three decades later?

Moreover, specialized hardware includes Xeon Platinum CPUs and a Mellanox-based communication system for the selected system, and the heterogeneity of hardware adds complexity in the analyses.

At this point, we must consider whether we can assess the carbon footprint of different systems, emphasizing the need for a methodology that accounts for both the manufacture and use of hardware and the construction of the building hosting the computer systems.

Additionally, it is imperative to examine macro- and micro-perspectives. A macro-perspective entails analyzing the carbon footprint across the entire lifespan of large-scale systems and correlating temporal changes to an ecologically acceptable trajectory by the year 2030. Meanwhile, a micro-perspective focuses on contemporary utilization of large-scale systems to foster new

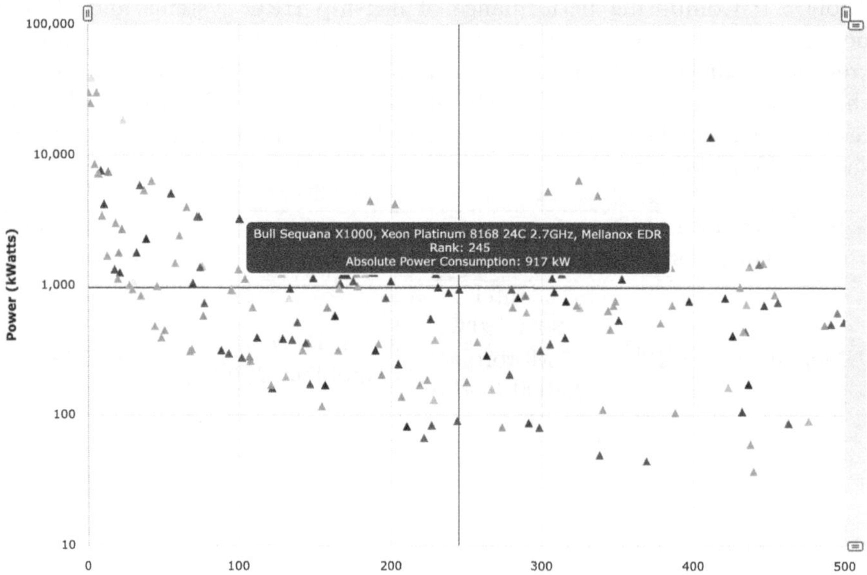

Figure 8.1: Power of the first 500 machines in the TOP500 list (Nov. 2024). Although significant effort toward energy efficiency was made for some top-10 ranked machines (x axis), they are among the most power-demanding machines (note the log scale on y axis).

practices at the end-user level. This approach may involve scheduling tasks during off-peak periods or devising schedulers that continuously account for the resources genuinely required by the application. This last point implies greater agility than today's static resource configuration decisions. We need more than one value to assess performance and track carbon emissions and impacts.

8.2.3 GREEN500

The idea of *GREEN500* was introduced in 2009, and the official ranking published on the TOP500 website began in 2013. It is now updated at the same rate as TOP500. It uses the same LINPACK benchmark, but it gives more attention to the energy efficiency (number of Floating Point Operations –FLOPs– executed per watt). GREEN500 contains the same attributes as TOP500, namely the hardware types, vendors, etc.

In 2011, Scogland et al. described the evolution of the GREEN500 metric over the first 3 years to identify the design aspects that contribute to estimating the impact of power consumption [82]. This study was extended in 2013 in [88]. It introduced a new (holistic) metric to give a unified view of the energy efficiency/performance of GREEN500. Gao et al. [36] have studied the influence of design and architectural parameters on the LINPACK scores in terms of performance and energy efficiency from TOP500 to GREEN500.

The correlation between performance and energy efficiency has been studied in Khan et al. [60]. They evaluated the evolution of Pearson's correlation coefficient for both lists over 10 years from 2009 to 2019, including the rise of GPU accelerators in HPC. They specifically focused on the comparison between homogeneous and heterogeneous systems in terms of performance and power consumption. Mair et al. [69] analyzed the energy efficiency of homogeneous and heterogeneous systems based on GREEN500 and proposed a new energy efficiency metric to avoid bias induced by system sizes. Hsu et al. mentioned the same metric in [50]. They provided a 10-year retrospective on the evolution of energy efficiency metrics and their exploitation to evaluate HPC systems (mainly the power usage efficiency and the performance-power ratio). Fraternali et al. [35] took a closer look at the impact of heterogeneity on the performance and variability of energy consumption. Although their study does not address trend analysis, they were able to propose some guidelines (hardware-wise) to build sustainable HPC systems.

In [61], the authors made projections of performance and energy consumption in 2011-2024, claiming that only heterogeneous systems will be able to reach exascale (10^{18} floating operations per second), but with a huge amount of energy up to 100 MW. Recently, effective exascale machines were obtained with El Capitan, Frontier, and Aurora systems, with a little less than 40 MW of power[1].

The report [6] offers a method for measuring and reporting power usage in HPC systems running workloads. This collaborative work with GREEN500, TOP500, Green Grid, and EEHPC WG focuses primarily on High Performance LINPACK (HPL), though it aims to benchmark various workloads. The document outlines four aspects of power measurement, each with requirements that become more stringent across three quality levels. The four aspects are the following:

1. Granularity, period, and type of raw measurement.
2. Machine fraction instrumented.
3. Subsystems included in instrumented power.
4. Location of measurement in the power distribution network, and accuracy of power meters.

8.2.4 Synthesis and social impact

In the realm of power or energy measurements, data is frequently collected from various locations, including, but not limited to, the output of the building

[1]`https://top500.org/lists/top500/list/2024/11/`

transformer. It is often necessary to account for power losses, and the specifics of these adjustments must be documented comprehensively in reports, such as those prepared for the TOP500 list. Ultimately, power or energy measurements should encompass all relevant factors.

Moreover, thanks to the 2025 report of the International Energy Agency [54], we now know that the order of magnitude of the power available in the datacenters in the 2030 horizon is between 1 GW and 10 GW.

Although table 2.2 in [54] characterizes the objectives for achieving total reliance on renewable energy sources, it is essential to critically assess whether this truly offers ethical reassurance. This consideration necessitates an exploration of the material implications associated with digital technology, a subject that will be further addressed in forthcoming discussions.

8.3 Tools for Measuring and Estimating Sustainability

8.3.1 Measurements

In the context of large-scale computing systems, arguably the main indicator of the use of natural resources is energy consumption [27, 43]. Possible explanations are, first, that energy consumption is viewed as the consumed electricity, which is an easily quantifiable metric. We can measure the electricity consumption of computing infrastructures (e.g., datacenters, supercomputers, computing nodes) at a high precision level, making it a suitable metric for optimization. Second, optimizing electricity consumption can be motivated by economic factors instead of factors related to sustainability/environmental impacts. An optimal electricity consumption leads to an optimal electricity bill.

Regarding computing nodes, Jay et al. [56] compared several tools to measure energy consumption. In summary, there are mainly two ways of doing such measurements: (1) measuring using dedicated hardware, and (2) measuring by using software. For the former, it can be further categorized by external devices and intra-node devices. External devices are types of equipment not attached inside the computing nodes, such as external power meters or Power Distribution Units (PDUs) that can measure power. Intra-node devices are equipments that are attached inside the computing nodes, such as Baseboard Management Controllers (BMCs). Both external and intra-node devices require extra hardware and thus extra materiality, which may outweigh their benefits in terms of sustainability.

For the latter, it uses sensors that are embedded into the Printed Circuit Board (PCB) of the computing nodes and software interfaces to measure the energy consumption. Software interface examples are the Intel Running Average Power Limit (RAPL) interface [26], and NVIDIA Management Library (NVML) [75]. Numerous works have built upon these interfaces to create energy measurement software tools [53, 92, 94], each having different difficulty levels to be used [56].

8.3.2 Estimations

Greenhouse gas emissions (in CO_2 equivalent emissions, hereafter referred to as CO_2 emissions) related to electricity usage are not measured directly. They are estimated by considering the carbon intensity of the electricity mix, which is a measure of CO_2 emissions per kilowatt hour of consumed electricity, as a proxy for estimating the emissions. Initiatives such as RTE[2] and ElectricityMaps (see Chapter 5) provide monitoring information about the electricity mix and its carbon intensity.

Regardless of the indicator (i.e., electricity consumption or CO_2 emissions), there are many other aspects regarding the consumption of natural resources that are not captured by measuring the electricity consumption. For instance, how many resources are used to manufacture the hardware? Related to this question, a recent report [77] estimated that around 60 % of datacenter impact in global warming comes from manufacturing. This share of global warming cannot be captured by looking at the amount of electricity that computing nodes consume. As such, several works [46, 101] have started considering the *embodied emissions*, which account for equipment manufacturing emissions.

8.3.3 Life-Cycle Assessment

Life-Cycle Assessment (LCA) is a methodology for evaluating the environmental impacts of a product, process or service. Its goal is to evaluate the impacts through the entire life-cycle, which includes raw material extraction, manufacturing, use and end-of-life phases. Typical LCA studies take into account the phases by quantifying their inputs in terms of natural resources (i.e., energy, water, metals, etc.) and the outputs in terms of environmental impacts (e.g., CO_2 emissions, water consumption, rare earth elements depletion). LCA is standardized by the ISO norm 14040:2006,[3] and is useful for identifying which parts of the product, process, or service bear the heaviest environmental burden. It is thanks to LCA that we can clarify, for instance, the impact of manufacturing computing equipment [77, 79].

Although LCA is a solid step toward accounting for environmental impacts, many challenges remain to be addressed. One of these challenges is on how to use LCA to account for large-scale computing services. We can bridge LCA studies related to single equipment (e.g., laptop, computing server, etc.) to extend to computing services. For instance, for a Cloud storage service, we can perform an LCA study on a functional unit related to deploying and maintaining a Cloud storage service for one year. Accounting for a

[2]https://www.rte-france.com/eco2mix/les-emissions-de-co2-par-kwh-produit-en-france

[3]https://www.iso.org/standard/37456.html

computing service requires Life-Cycle Inventory (LCI) data for all types of equipment used in the service supply chain. This includes (without being exhaustive) LCI data for storage servers used in production, laptops used for service development, servers used for development/testing, network infrastructure (routers, cables), etc. Such LCI data for all supply chain equipment may not always be available.

As an example, Berthelot et al. [12] used an LCA approach to evaluate the environmental impacts of an image generation service that uses the Stable Diffusion model. They considered two functional units: (1) one prompt in the service, generating four images, and (2) running the service for one year. Regarding equipment and supply chain, they considered notably the AI servers used for training and inference, end-user terminals, network connection, and the server used for service web hosting. On one hand, evaluating a service rather than individual equipment enlarges the evaluation boundary, which helps identify which parts of the service use more resources. On the other hand, computing services are often complex, which results in intricate supply chains with many particular steps, which hinders the evaluation with other similar services.

8.4 Green HPC

The general message from this section is that the computing notion is evolving, both technically and in terms of its social impact, and it is becoming hidden under the notion of Cloud. However, computing comes at a cost. Making it more visible should improve the accountability of those using it and, therefore, the uses made of it.

Here, we are interested in the notion of "Green HPC," that is, the means of building computing systems (hardware and software) that have the advantage of being more sustainable than in the past. In short, the community has made significant progress in hardware (compute nodes, accelerators, memory, interconnects) that increases energy efficiency. However, it is shown in [11] that the progress in performance is greater than that in energy efficiency.

Finally, we also provide additional and complementary material to Chapter 4, specifically for Subsections 4.3.3 and 4.3.4.

8.4.1 Hardware and infrastructures

According to Table 8.1, the performance (flops/s) per watt has consistently increased over 40 years, supporting the idea that HPC is environmentally friendly. Power Usage Effectiveness (PUE) measures datacenter energy efficiency, calculated by dividing total power by IT equipment power. A lower

PUE, tending toward 1.0, indicates better efficiency. Most datacenters have a PUE just above one, highlighting sustainability.

The PUE metric is designed for repeated use to assess datacenter efficiency changes. It helps identify effective and ineffective energy reduction methods. Organizations with good PUE scores can leverage them in marketing, but the lack of a standardized measuring methodology poses limitations.

Cloud Continuum represents the expanding adoption and innovation of cloud computing, its diverse applications, and the agility it provides. AI has significantly advanced in areas like speech recognition, visual perception, search, language translation, and sector-specific uses in manufacturing, education, and healthcare. IT intelligence is harnessed through automation to optimize infrastructure, enhance testing quality, and strengthen security mechanisms.

The hardware behind these innovations is the same as in HPC datacenters (CPU, GPU, NPU). The cloud service model allows HPC computing to be treated like any other service, such as a web server. Communities have embraced this to manage both finite HPC jobs (online service in Alibaba terminology) and internet jobs (batch jobs in Alibaba terminology). In 2014, Alibaba introduced the FUXI scheduler [100], a fault-tolerant resource management and job scheduling system at internet scale. Alibaba also offers the publicly available Alibaba Cluster Trace Program [5], which provides insights into internet datacenters and workload characteristics. A notable concept from Alibaba is co-location, where HPC and internet jobs run on the same machine.

Liu and Yu [66] highlight three findings distinct from the Google trace [1]: online services operate in containers, while batch jobs use physical servers, managed by different schedulers on the same servers. Batch tasks primarily utilize unused container resources, demonstrating the elasticity of Alibaba cluster resource utilization through overprovisioning and overcommitment. To prevent performance issues for online services, the cluster limits resources for batch tasks, ensuring a balance termed "resource plasticity".

Finally, let us briefly examine a classic problem: server consolidation. Server consolidation is an approach to efficiently use computer server resources to reduce the total number of servers when the load becomes small. It can also consist of increasing the number of servers when the load becomes high. In this case, we speak of dynamic scaling, and it is widely used in the context of the Cloud. Consolidation was discussed in Subsection 4.3.4, but not in the context presented here, which offers another opportunity to move toward sustainable solutions.

Consider that instead of the user requesting a fixed amount of CPU, RAM, and disk space, they request a range of values. They specify that they want between six and ten cores, rather than specifying eight cores. This abstract view allows the authors [71, 72, 73] to relax the placement problem and make better placements than in the case of ranges reduced to a single value.

8.4.2 HPC platforms and datacenters

As already mentioned, digital technology (ICT) in general and datacenters in particular have a growing impact on climate change, with CO_2 emissions growing year-over-year, accounting for around 4 % of the global CO_2 emissions [70]. However, considerable efforts have been made in terms of processor consumption, networks, and even cooling, with the recent introduction of an adiabatic cooling process based on water vaporization. As a result, the impact of datacenters is less than it would have been without these efficiency gains.

With the booming growth of AI technology, datacenters have started becoming closer to HPC platforms. For instance, current large generative AI models require the order of hundreds of gigabytes of memory to be hosted, and accelerator devices such as Graphics Processing Units (GPUs), Tensor Processing Units (TPUs), or Application Specific Integrated Circuits (ASICs) to run the models. We can therefore grasp the future dynamics of HPC platforms by taking the growth of high-performance AI datacenters as an example.

Shehabi et al. [83] presented estimations and future scenarios for the datacenter electricity consumption in the United States. They showed that, to meet AI workload demand, the datacenter share of the total electricity consumption in the United States will jump from around 4 % in 2023 to 12 % in 2028 following the tendency scenario.

This AI workload demand nullifies all previous efforts to reduce energy consumption by increasing energy efficiency. Beyond energy consumption, entire datacenter parks have to be renewed, opting for the fastest and most energy-efficient servers, which means a huge consumption of metals and rare-earth elements. Moreover, liquid-based cooling processes aiming to increase the PUE [90] result in an increased water consumption [42, 98], another ecological impact of the growing use of datacenters.

8.5 Management of Renewable Resources

Leaving aside the environmental impact of large-scale computing infrastructures and datacenters due to their manufacturing, the main disadvantages linked to the usage stage are firstly the carbon-based electricity used to supply the platforms, and secondly the impact of such a huge energy consumption on sharing this energy in all economic sectors, low-carbon industry, transport and also private individuals.

The question here is to present and discuss alternative solutions for carbon-free computing where energy is potentially produced on-site. The carbon impact of server manufacturing is not taken into account here.

8.5.1 Renewable energy sources for ICT (HPC and DC)

Despite the progress made in increasing the energy efficiency of hardware components in datacenters [13, 37], including facilities such as the cooling system [21, 99], of which there are many variants [32], their impact on greenhouse gas emissions is increasing as datacenters become more numerous [89] and their uses multiply, in part with the growing use of AI [54] in general and generative AI in particular [65]. If we focus on the use phase, the negative impact of datacenters is their carbon-based electricity consumption, which is responsible for CO_2 emissions on behalf of datacenters. For other negative ecological impacts, such as the consumption of water to the cooling process of servers, it is possible to find a geographic location that allows free cooling [99] as Facebook has done by installing datacenters close to the Arctic Circle [45], or Microsoft by submerging a datacenter underwater in the Atlantic Ocean near the north of Scotland [51]. However, this is not always possible, and concentrating all the datacenters at the North Pole does not make sense. If we want to reduce the datacenter's water consumption, the cooling system has to be based only on electricity, thereby increasing the datacenter consumption and therefore CO_2 emissions. Another possibility to explore is to decarbonize datacenter electricity as much as possible. In this case, either we have to resort to electricity generation using nuclear power plants, with all the risks that this entails, or we have to produce this electricity using only renewable energy [78].

Providers have already explored this option offsetting their GHG emissions by purchasing clean energy on the other side of the country [8, 15, 44]. Since 2017, Google operates its datacenters using 100 % renewable energy, but not generated on site [52]. The deal is to reinject the same amount of electricity back into the country's power grid as is being consumed at the same time, using electricity generated from renewable energy sources such as wind and sun, precisely where there is sun or wind or both, while it may even be night time where the datacenter is consuming electricity. While green energy is controllable, such as electricity generated using hydropower, the datacenter or HPC platform is not quite green all the same. In fact, the use of a huge quantity of electricity on the power grid can unbalance the grid, forcing the electricity supplier to re-inject the equivalent in brown electricity to compensate for this use, as they have no other controllable sources at their disposal, and thus enable other users to have access to electricity [23]. This means that offsetting will never compensate for the CO_2 emitted to satisfy the datacenter, as the CO_2 emitted is a virtually irreversible operation [18, 93]. The datacenter consumption must therefore be removed from the circle of other consumers to avoid generating this bias [85].

Another approach is to generate and store electricity from renewable energy sources on the datacenter site itself. Pierson et al. [78] have demonstrated that supplying a mid-sized datacenter (1 MW) solely with renewable energy (sun and wind) entirely generated on site is feasible from both a management and a sizing point of view [47]. However, if we take the economic aspect into account, using electricity from the grid remains the cheapest solution in most countries. Autonomous configurations are also more expensive than solutions allowing access to the grid, where electricity can be reinjected or taken back according to the needs of the datacenter. Finally, a stand-alone solution with hydrogen storage is preferable to battery-only storage [41]. It is easy to understand that the intermittency and uncertain character of renewable energy sources are a major issue when it comes to low-carbon energy. This requires additional resources and/or mitigation measures, and IT companies, rather than dealing with this problem themselves, might prefer to use renewable energy sources already on the power grid. Once again, with this approach, there is no guarantee that the electrons supplying the datacenter are actually "green," as, due to the laws of physics, they will typically originate from the nearest plants, green or not.

So, to guarantee the low-carbon origin of its electricity, it is preferable to produce on site from renewable energy sources to supply all or part of the electricity needs of the datacenter [80]. But this energy mix poses many technical and economic challenges, such as energy storage [19]. While cost estimates have significantly decreased over time [64], they are still considered as prohibitive for many applications [20]. However, a mixed solution with both local production and grid connection has its advantages. Indeed, grid-connected datacenters can become auxiliary services to the grid [2, 4, 39], i.e., they can absorb peaks or provide short-term energy and thus participate in frequency regulation or reserves. This behavior can be achieved either by modulating the IT load [7], or by using local storage resources [24]. This can generate an additional revenue stream for datacenter operators, as these services can be sold on the market but require the use of portfolio optimization techniques [38].

8.5.2 Resource management of intermittent and heterogeneous resources

As soon as the datacenter is powered by renewable energy sources, even partially, or as soon as we take into account the carbon content of the electricity available on the grid, the envelope of power that is as virtuous as possible to power the datacenter can no longer be considered constant over time. In addition, the workload of the IT part of the datacenter is not constant either [25], nor is there any reason why the datacenter energy requirements should coincide perfectly with this energy envelope. A negotiation is therefore necessary between power demand and energy supply, to ensure that the datacenter carbon footprint is as low as possible during the usage phase [78, 91]. To achieve this goal, several levers can be activated: shifting the workload of the datacenter [59], and using energy storage [48]. This resource management policy involves solving optimization problems, most of which are combinatorial and difficult to solve. These problems classically belong to the family of scheduling problems, which can be addressed using mixed integer linear programs, or heuristic approaches for the most difficult ones [59].

To a first approximation, the inputs to this management process are time series providing workload prediction, modeling initial electricity demand, weather conditions to predict primary renewable energy production, and energy levels stored in batteries, hydrogen systems or any other storage devices that can be activated for charging or discharging. The results of the optimization problem, solved globally or separately, consist of time series which are also the setpoints describing the number of machines to be activated, time slot by time slot, as well as the charging or discharging power of all available storage devices to contribute to the datacenter supply process, also time slot by time slot under the negotiated conditions. These optimization steps are offline optimization processes [16, 63].

Given the uncertainties associated with the input data for the problems described above, compliance with these setpoints is not a robust approach. If the wind is not as strong as expected, primary energy production is not at the expected level, and to guarantee the same power to supply the datacenter at that time, a storage device must discharge part of its stored energy to compensate for this missing energy production. An online process therefore needs to be implemented to take account of actual conditions, workload (e.g., the actual time taken to complete a submitted task, which may differ from the time initially planned), weather conditions (e.g., wind strength or sunshine not as abundant as expected), the efficiency of any device, which may depend on the ageing and health status of platform components, etc. An online approach may be that of automation processes, which are capable of knowing the extent to which it is possible to deviate from setpoints [29]. Decisions can then take into account actual conditions and predicted values for the near future, for instance, switching servers off or on, migrating tasks, charging or discharging more or less energy from short- or long-term storage devices (batteries of hydrogen systems) [28]. Figure 8.2 illustrates four scenarios in which over- or under-power production by renewable energy sources, which are intermittent by nature, is compensated by storage devices. The management of these long- or short-term storage devices is a key point in the negotiation process to guarantee the given quality of service.

8.5.3 The question of regulation

We discussed several ways of reducing the carbon emissions of computing platforms, namely, energy efficiency (increasing the number of operations per watt), and changing the power supply to on-site renewable energy. Even so, the overall energy consumption of the sector continues to rise, but the global available energy envelope is finite, and more and more companies, administrations, and private customers aim at increasing their power demands. It remains a lever to study: regulating the usage stage in the fairest possible way.

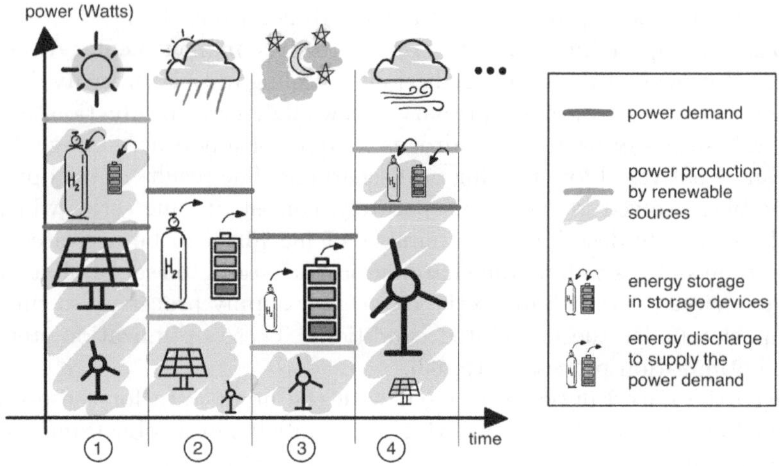

Figure 8.2: Four scenarios in which storage devices mitigate the inherently intermittent production of renewable energy by charging or discharging batteries or hydrogen reserves so that the DC power supply is adequate: (1) sunny day; (2) rainy day; (3) windy night; (4) windy day.

Regulation is the set of rules governing access to a shared and limited resource. This regulation can either come from above, via a centralized regulatory body (by the legislator or by the market), or from below, via decentralized decisions made by the users themselves, either out of interest (incentive policies) or conviction (user communities). Both aspects are discussed below.

From a technical perspective, it is always possible to imagine a power generation infrastructure as part of a smart grid that would feed the non-consumed surplus electricity by datacenters back into the global (national or regional) power network, and take back some of the green electricity produced by other renewable energy producers in case of local energy shortage, if possible. This would mitigate the intermittency of renewable energy sources and limit the oversizing required to feed storage facilities for use when there is no sun or wind. In the best case, datacenters appear both as the problem and the solution. But, such a solution is not easy to design (mainly in billing policies and governance). This will probably not be enough if we are in a regime where the share of renewable energy is too high. Moreover, it will require a change in usage for users to adapt and accept [74].

From a social perspective, focusing on system managers and private consumers, the main concern is about developing efficient incentive mechanisms. Such mechanisms can be economical or ethical (reward mechanism) [3, 67].

Engaging and raising awareness among users about the responsibility of datacenters to share electricity with society is almost nonexistent in current computing platforms.

8.6 Job Resource Management

8.6.1 Context

Computing platforms consume more and more power, reaching levels [76] close to the power consumption of a small city. Such a level of power consumption may clash and cause tension in the electricity grid, especially in extreme events. Extreme events may be, for instance, climate events that induce stress in the electricity grid, such as the 2021 Texas power crisis [95].

It would be irresponsible to operate computing platforms at full power in the above situation. Further stressing the electricity grid can result in a full grid collapse. Alleviating this stress involves giving answers to the following question: *How can we effectively reduce the platform power consumption while keeping it operational?* We recommend informing and interacting with the platform users and administrators to effectively address this question. We give more details below.

8.6.2 Data and analytics

Many users submit parallel applications (i.e., jobs) to run on a computing platform. Job submissions are managed by a Resources and Jobs Management System (RJMS), with Slurm [97] arguably being the most popular RJMS. The RJMS typically logs metadata regarding the submitted jobs, such as their arrival time, the requested number of nodes/processors, and an estimation of their processing time. Jobs' allocation and execution are usually represented in a Gantt chart (top graph in Figure 8.3).

We can enrich this metadata to give us an idea of how much the jobs draw power over time (i.e., energy consumption). To obtain this data, recent HPC platforms are deployed with energy consumption monitoring tools (see Section 8.3.1). These tools can provide power consumption data at the computing node level, as a time series of the power consumption of the computing node (bottom graph in Figure 8.3).

We can merge the RJMS job data with the computing node power monitoring data to get an idea of job power consumption. In the case of jobs that share a same computing node (stacked rectangles in Figure 8.3), the resulting power consumption is in function of each of the job energy consumption plus potential interference between the jobs, which makes it hard to distinguish each job's contribution to the node power consumption. When taking into account jobs that have exclusive access to computing nodes, however, we can identify a power consumption profile of the jobs. The exact power profile of the job can only be known after the job is performed.

8.6.3 Conduct systemic changes rather than just optimize

Knowing how much power jobs draw over time (i.e., energy consumption) gives us a new dimension for optimization. Several works have explored this new dimension, proposing algorithms and scheduling policies aiming at optimizing the power usage and the scheduling efficiency, as discussed in Chapter 3. Although significant progress has been made in energy-aware scheduling research, it is hard to translate theoretical gains into practice for reasons that boil down to how users and administrators work and interact with the computing platform.

Each computing platform has its own ecosystem, with specific cultural and organizational behaviors. Users can, for instance, "work around" the scheduling policies in such a way that suits them, which may lead to a negative synergy with policy goals. Scheduling algorithms or policies can be even rejected [34] by HPC platform maintainers/users because they are either (1) impractical to be deployed and maintained, or (2) the reasoning behind their scheduling decisions is too hard to be explained.

We believe that transitioning toward a more sustainable job resource management requires knowing the specific cultural and organizational behaviors of the people who use and work with the computing platform, rather than just employing technical optimizations. Understanding these specific cultural and organizational behaviors can open the door for energy reductions that otherwise would not be visible. For instance, Giordani et al. [40] observed

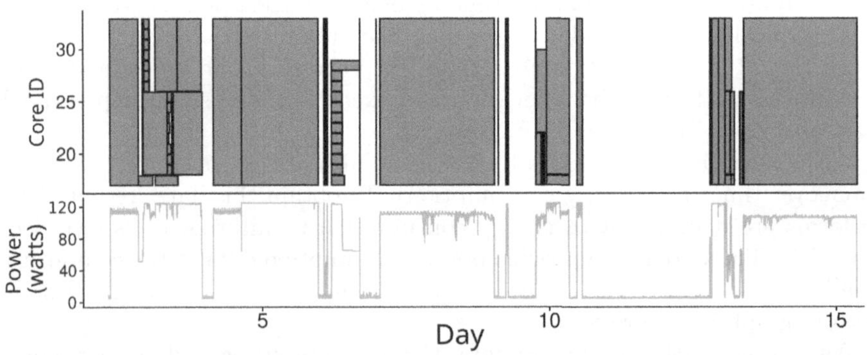

Figure 8.3: An example of two sources of trace data: (top) RJMS execution time and allocation data (processing time vs. cores allocated), and (bottom) power monitoring of the computing node (power vs. time). Figure adapted from [17].

that, in their use case, 40 % of users may plan their schedules around the processing time of their applications, having less than 5 % of active waiting time for the applications to finish. Put another way, any optimization technique (including scheduling algorithms and policies) to run the applications faster may not be perceived by 40 % of the users, who are possibly at maximum working bandwidth. Moreover, they showed that around 50 % of the studied users would perceive only a minor impact in their work if the performance of their applications increased 10-fold. The same logic could be applied to run the applications more slowly or to finish them later to save energy or reduce tension in the electricity grid. Some users may perceive only minor impacts (or even no impact at all) if their applications run slightly slower or if they finish later, because these users can schedule their work time around the application processing time. We can leverage this knowledge with job energy consumption and processing time statistics (Section 8.6.2) to make systemic changes in the platform ecosystem that result in higher-order positive effects (Section 8.8.2). For instance, we could give incentive mechanisms [84] to run applications in low power-stressed situations, and we could inform and instruct users on how to better schedule their work around the application processing time, in such a way that running the applications slower or finishing them later to save power/energy does not result in impacts in their work.

8.7 HPC for Green

The computing demands are always increasing [14]. Section 8.2 highlighted the footprint of HPC systems using data from TOP500 and GREEN500, assuming these systems are representative of the whole HPC domain. Key points include the HPC domain growth at a rate near Moore's law (floating point operations per second), and that GREEN500 energy gains are less than performance gains. Koomey's law [62] is equivalent to Moore's law for the evolution of the number of floating operations per watt. This rate increases almost linearly on average for machines on the TOP500 list.

However, if we consider maximum efficiency on this same list, this efficiency stagnates and has not increased since mid-2022. HPC applications require significant resources and efforts, but should reduce GHG emissions. In addition, to develop scientific applications that require high performance, developers have to be aware of the last innovations in the whole HPC domain including hardware, power management, architecture design, scheduling policies, etc., to take advantage of the last technologies to reduce the energy consumption of their HPC applications [87]. Since that does not seem enough, the question is: "To what extent does HPC help mitigate global warming?"

Researchers and decision-makers see computing, and in particular HPC, as a potential solution to the environmental crisis. For instance, large-scale climate simulations are essential for understanding global dynamics. Many researchers, including [10], imagine that AI could replace simulation calculations with inference. However, developing AI learning components remains costly in terms of time and computing power [54], even if there exist tools to develop an AI more environmentally friendly and inclusive [81]. Given all roles that machine learning can play, AI may have a positive or a negative impact on GHG emissions depending on the domain [22, 58].

A point of vigilance must be observed when it comes to AI contribution to solving the problem of global warming, as AI can also appear to be part of the problem. In July 2024, Google announced in the blog, *Our 2024 Environmental Report* [44], that: "In 2023, our total GHG emissions increased 13 % year-over-year, primarily driven by increased datacenter energy consumption and supply chain emissions". AI is pulling the digital domain upward (leading to higher consumption and getting out of control, which raises ethical questions that we will not detail here. The virtuous approach of building an exascale HPC system while limiting consumption to 20 MW no longer exists.

8.8 Beyond Efficiency: Sufficiency

In the previous sections, we presented several ways for significantly reducing the impact of computing platforms. Multiple gains are possible for particular applications running on particular platforms, at both hardware and software levels. Even when these applications aim to design a service that reduces carbon emissions, quantifying these gains precisely is very difficult. We need to be able to estimate the entire computing environment, not just the electrical cost of running the application. The overall balance must be positive, and there are only a few applications whose overall impact has been demonstrated to be truly positive. For instance, Section 8.4 described an approach for building and exploiting greener HPC and DC platforms, namely, running on more energy-efficient hardware (or with more renewable electricity) and improving the job management, mainly by taking into account the energy consumption as a primary objective instead of pure performance. Another way is to cap the power in critical periods of brown energy or to delay some jobs in order to improve the overall energy consumption. It is possible to analyze the footprint of a computing platform using classical attributive LCA, plus complementary analyses of the part of electronic devices used. "Attributive" means that LCA is restricted to the evaluation of components directly attributed to the service. Of course, this approach is useful for reducing existing negative impacts, but it does not fundamentally call anything into question, and thus, the estimation of the gains is limited compared to the growth of the field. This approach is called *efficiency*, which can be summarized as doing the same with less. The biggest problem here is the embalming effect of the entire domain, i.e., what digital technology enables as an acceleration of other domains. For instance, one can significantly improve the cost of oil exploration by using large-scale numerical methods running on HPC platforms, but in return, oil companies will extract more oil (more efficiently, faster, and at a better price) to meet society's ever-increasing demand. The result is more CO_2 emissions. In this section, we show how we can go beyond the efficiency approach to take usage into account. This is called *sufficiency*, and it corresponds to doing less (if possible in an efficient way).

8.8.1 Logic of systemic effects – From single applications to global effects

We already discussed positive environmental effects of HPC (Section 8.7). Let us clarify the hidden part. An application could have positive effects if it serves to avoid emissions in other sectors. However, even in this case, it may lead to negative impacts. Let us consider, for instance, a geophysics application, designed for climate and Earth-system modeling. It is positive if it consists of simulations in CO_2 capturing or in identifying methane fleets, but it can also be used as a basis for oil and gas exploration that will lead to more GHG emissions after extraction. To what extent can we assess *in fine* that a given application guarantees positive impacts? Are we sure that it is globally positive? The previous example emphasizes a complex mechanism that is usually not considered in direct footprint analyses, but whose impacts are potentially strongly negative. The footprint is not only to be considered at the level of a single application or at the level of a single platform. It should include the effects outside the strict perimeter of the applications, for instance, a new improved numerical method that solves a problem, possibly more efficiently, will be used for solving other similar problems, leading to more energy consumption. Figure 8.4 illustrates this acceleration principle.

It is very difficult to capture the systemic effects since classical analyses like LCA are restricted to specific evaluations (attributive LCA), they do not consider how the applications are used and what new features or variants they contribute to developing. Another difficulty is that it is impossible to evaluate all the carbon emissions saved, whatever the domain (transport, buildings, food production, logistics), using HPC modeling and simulations, as we already discussed in this chapter.

The technological gains (like more energy-efficient hardware or improved resource management) are often overcompensated by either executing more runs of the same application, running more applications, or even installing larger computing and storage systems for the same monetary costs, such that the overall environmental impact is not reduced. This is a well-identified mechanism, generally referred to as the Jevons paradox or rebound effect [57], which can only be addressed at a larger scale by decision-makers (i.e., funding agencies or governments) setting up strict upper limits on the overall consumption [68].

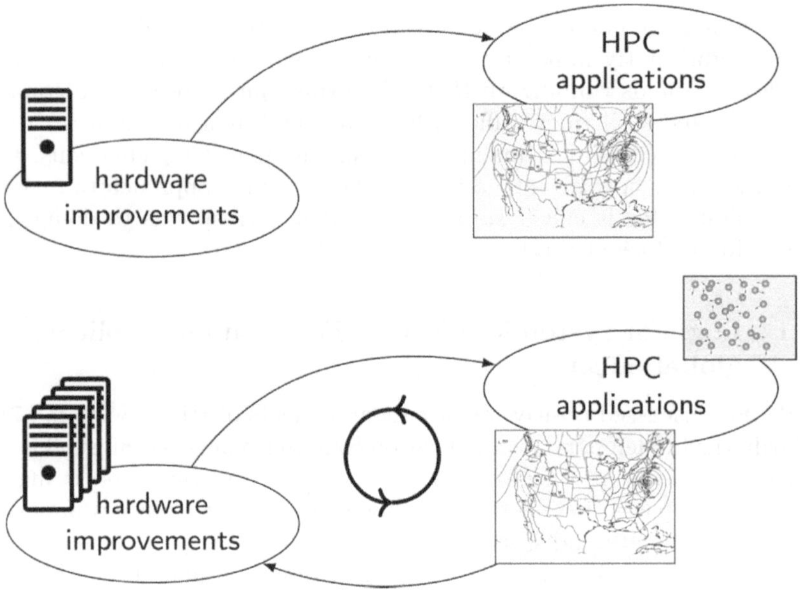

Figure 8.4: Acceleration dynamics: at the top, powerful hardware allows large and efficient applications to run (for instance, climate simulation); at the bottom, in turn, applications evolve and more features are added (like interactive visualization), better accuracy is targeted, other variants of the original problem are developed, etc. Such new usages need more efficient hardware that allows new applications to be developed, and so on.

However, a few rules of good conduct can be laid down to limit these upstream effects. Some natural recommendations are to inform developers of applications, to involve decision-makers, and to anticipate by putting regulation mechanisms to limit potential counterproductive extensions. We have discussed regulation in Section 8.5.3, whose main challenge is to bridge the gap between the microscopic level of users and the macroscopic level of the entire domain. Both parts need to be considered together.

8.8.2 Accounting for higher-order effects

Hilty et al. [49] defined ICT's second-order effects as the "indirect environmental effects of ICT due to its power to change processes (such as production or transport processes), resulting in a modification (decrease or increase) of their environmental impacts" [49]. In turn, higher-order effects are indirect impacts that occur as a consequence of the second-order effects and often occur following behavioral or structural changes, such as changes in consumption patterns, lifestyles, and value systems [33]. Regarding LCA, most studies fall in attributional LCA, where second- and higher-order impacts are not considered. A broader approach is to consider *consequential* LCA, which attempts to respond to the limitations of attributional LCA by identifying the most impactful possible consequences, which can at least qualitatively be taken into account. Yet, there is little to no literature on how to bridge consequential LCA (or second- and higher-order effects in general) and large-scale computing systems. Building this bridge requires audit methods that link quantitative indicators with qualitative indicators that capture the specific systemic interactions with the large-scale computing system and its "environment" (e.g., users and society). Let us mention that a non-exhaustive solution is to complement the classical footprint analyses with macroscopic indicators, like, for instance, the global market of HPC vendors.

To put it into perspective, Chapter 5 presents the GHG protocol classification of companies' emissions in scopes 1, 2, and 3. While this scope classification is a well-established standard for accounting emissions, it fails to account for the *consequences* (i.e., second- and higher-order effects) of deploying a digital service or solution. The scope 4, coined by the World Resources Institute as the "emission reductions that occur outside of a product's life cycle or value chain, but as a result of the use of that product" [96] attempts to account for the consequences, but it presents a major drawback because it accounts only for reductions, which we arguably say that is not the case in the majority of digital services and solutions. Moreover, rebound effects do not fit into any of the scopes (1 to 4).

8.9 Conclusion

In this chapter, we have studied the environmental impacts of large-scale computing systems. We have focused our analysis on the execution of applications on computing platforms, as measured by the amount of electricity consumed at runtime. However, significant negative impacts must be added to this phase of use, such as the materiality of electronic components (CPUs, accelerators, memory, interconnection networks, and, more broadly, the impact of building and using premises). These extra environmental impacts are measured in particular by the quantity of water and metals used for building and assembling integrated circuits. Note here that many HPC systems and DCs belong to private companies that do not communicate their related data, making such a global analysis very difficult.

The question of how the HPC and DC domains can benefit the climate crisis is complex. Of course, many applications are developed for a better understanding of climate dynamics or for optimizing codes that reduce heavy environmental impacts (which become always more important, for instance, with the rise of AI).

We have shown how we can take actions to manage resources in a way that takes carbon emissions into account. This is an important direction, well anchored in the traditional skills of ICT specialists, which will not allow significant optimization of electricity consumption. It is not up to the challenges of climate change that need fast and deep reactions. Another direction is to build renewable platforms, which also poses new scientific challenges in optimizing heterogeneous and intermittent resources. Here, however, it has to be said that the electricity produced by renewable sources of energy should be shared with local populations. The power needed by a medium-sized DC accounts for 100,000 people.

HPC can also be used as a lever to mitigate carbon emissions and other negative effects of planetary limits, but this is almost impossible to measure. What is certain is that the use of HPC also contributes to more carbon emissions, and that a positive overall balance must be guaranteed by counting all impacts. This is possible for a given application running on a specific platform, but we cannot simply add the contributions of all the applications since there are some common shares. Another problem is to identify and estimate the indirect and higher-order effects induced by the use of HPC in other domains. We presented in this chapter the basis to help people understand the problem as a whole, since we need to do more than just develop efficient applications: such an endeavor does not directly reflect the overall picture.

Many researchers are aware of the problems, but they do not really know how to go further than technical solutions like energy-efficient applications and system management. The next step toward green HPC remains long and challenging since we need to move on to a broader economic, sociological, philosophical, and political perspective. A purely scientific view of the problem is insufficient to take into account the impacts of usage. This is known as the Collingridge dilemma: Until a technical solution has been fully developed, we do not know whether it will really bring any benefits, and once it has been developed, it is too late to go back and include the usage dimension.

8.10 References

[1] O. A. Abdul-Rahman and K. Aida. Towards understanding the usage behavior of google cloud users: The mice and elephants phenomenon. In *Proceedings of the 2014 IEEE 6th International Conference on Cloud Computing Technology and Science*, CLOUDCOM '14, page 272–277, USA, 2014. IEEE Computer Society.

[2] D. Aikema, R. Simmonds, and H. Zareipour. Delivering ancillary services with data centres. *Sustainable Computing: Informatics and*

Systems, 3(3):172 – 182, 2013. Selected papers from the 2012 IEEE International Green Computing Conference (IGCC 2012).

[3] C. Akasiadis, K. Panagidi, N. Panagiotou, P. Sernani, A. Morton, I. A. Vetsikas, L. Mavrouli, and K. Goutsias. Incentives for rescheduling residential electricity consumption to promote renewable energy usage. In *2015 SAI intelligent systems conference (IntelliSys)*, pages 328–337. IEEE, 2015.

[4] S. M. Ali, M. Jawad, M. U. S. Khan, K. Bilal, J. Glower, S. C. Smith, S. U. Khan, K. Li, and A. Y. Zomaya. An ancillary services model for data centers and power systems. *IEEE Transactions on Cloud Computing*, pages 1–1, 2018.

[5] Alibaba Group. Alibaba cluster trace program. `https://github.com/alibaba/clusterdata`. Accessed: (April 2025).

[6] Anonymous. Energy Efficient High Performance Computing Power Measurement Methodology. `https://github.com/EEHPCWG/PowerMeasurementMethodology`, 2013-2025. Accessed: (April 2025).

[7] M. Antal, C. Pop, D. Valea, T. Cioara, I. Anghel, and I. Salomie. Optimizing data centres operation to provide ancillary services on-demand. In J. Altmann, G. C. Silaghi, and O. F. Rana, editors, *Economics of Grids, Clouds, Systems, and Services*, pages 133–146, Cham, 2016. Springer International Publishing.

[8] Apple. Supplier Clean Energy 2022 - Program Update. `https://www.apple.com/la/environment/pdf/Apple\_Supplier\_Clean\_Energy\_Program\_Update\_2022.pdf`, 2022. Accessed: (April 2025).

[9] C. Belady. In the Data Center, Power and Cooling Costs More than the IT Equipment It Supports. *Electron Cool*, 13, 2007.

[10] Z. Ben-Bouallegue, M. C. A. Clare, L. Magnusson, E. Gascon, M. Maier-Gerber, M. Janousek, M. Rodwell, F. Pinault, J. S. Dramsch, S. T. K. Lang, B. Raoult, F. Rabier, M. Chevallier, I. Sandu, P. Dueben, M. Chantry, and F. Pappenberger. The rise of data-driven weather forecasting, 2023. `https://arxiv.org/abs/2307.10128`.

[11] A. Benhari, D. Trystram, F. Dufossé, Y. Denneulin, and F. Desprez. Green HPC: An analysis of the domain based on Top500. *arXiv preprint arXiv:2403.17466*, 2024.

[12] A. Berthelot, E. Caron, M. Jay, and L. Lefèvre. Estimating the environmental impact of generative-ai services using an lca-based methodology. *Procedia CIRP*, 122:707–712, 2024.

[13] K. Blok. Improving energy efficiency by five percent and more per year? *Journal of Industrial Ecology*, 8(4):87–99, 2004.

[14] A. Borghesi, A. Bartolini, M. Lombardi, M. Milano, and L. Benini. Predictive modeling for job power consumption in HPC systems. In *High Performance Computing: 31st International Conference, ISC High Performance 2016, Frankfurt, Germany, June 19-23, 2016, Proceedings*, pages 181–199. Springer, 2016.

[15] J. Calma. Google's future data centers will be built next to solar and wind farms, Dec. 2024. `https://www.theverge.com/2024/12/10/24317888/googles-data-center-solar-wind-renewable-energy-intersect-partnership`. Accessed: (April 2025).

[16] L.-C. Canon, D. Landré, L. Philippe, J.-M. Pierson, and P. Renaud-Goud. Assessing power needs to run a workload with quality of service on green datacenters. In *European Conference on Parallel Processing*, pages 229–242. Springer, 2023.

[17] D. Carastan-Santos, G. Da Costa, M. Poquet, P. Stolf, and D. Trystram. Light-weight prediction for improving energy consumption in HPC platforms. In *European Conference on Parallel Processing*, pages 152–165. Springer, 2024.

[18] Carbone4. Stop saying "carbon offset": Moving from "offsetting" to "contributing", July 2019. `https://www.carbone4.com/stop-saying-carbon-offset-from-offsetting-to-contributing`. Accessed: (April 2025).

[19] A. Castillo and D. F. Gayme. Grid-scale energy storage applications in renewable energy integration: A survey. *Energy Conversion and Management*, 87:885 – 894, 2014.

[20] B. Celik, S. Suryanarayanan, R. Roche, and T. M. Hansen. Quantifying the impact of solar photovoltaic and energy storage assets on the performance of a residential energy aggregator. *IEEE Transactions on Sustainable Energy*, 11(1):405–414, Jan. 2020.

[21] J. Cho and Y. Kim. Improving energy efficiency of dedicated cooling system and its contribution towards meeting an energy-optimized data center. *Applied Energy*, 165:967–982, 2016.

[22] P. Clutton-Brock, D. Rolnick, P. L. Donti, and L. Kaack. Climate change and AI: Recommendations for government action. Technical report, GPAI, Climate Change AI, Centre for AI & Climate, 2021.

[23] A. Coskun. AI supercharges data center energy use – straining the grid and slowing sustainability efforts. The conversation, July 2024. `https://theconversation.com/ai-supercharges-data-center-energy-use-straining-the-grid-and-slowing-sustainability-efforts-232697`. Accessed: (April 2025).

[24] L. Cupelli, N. Barve, and A. Monti. Optimal sizing of data center battery energy storage system for provision of frequency containment reserve. In *IECON 2017-43rd Annual Conference of the IEEE Industrial Electronics Society*, pages 7185–7190. IEEE, 2017.

[25] G. Da Costa, L. Grange, and I. de Courchelle. Modeling, classifying and generating large-scale google-like workload. *Sustainable Computing: Informatics and Systems*, 19:305–314, 2018.

[26] H. David, E. Gorbatov, U. R. Hanebutte, R. Khanna, and C. Le. Rapl: Memory power estimation and capping. In *Proceedings of the 16th ACM/IEEE international symposium on Low power electronics and design*, pages 189–194, 2010.

[27] M. Dayarathna, Y. Wen, and R. Fan. Data center energy consumption modeling: A survey. *IEEE Communications surveys & tutorials*, 18(1):732–794, 2015.

[28] I. F. De Nardin, P. Stolf, and S. Caux. Analyzing power decisions in data center powered by renewable sources. In *2022 IEEE 34th International Symposium on Computer Architecture and High Performance Computing (SBAC-PAD)*, pages 305–314. IEEE, 2022.

[29] I. F. De Nardin, P. Stolf, and S. Caux. Mixing offline and online electrical decisions in data centers powered by renewable sources. In *IECON 2022–48th Annual Conference of the IEEE Industrial Electronics Society*, pages 1–6. IEEE, 2022.

[30] J. Dongarra, J. L. Martin, and J. Worlton. Computer benchmarking: Paths and pitfalls: The most popular way of rating computer performance can confuse as well as inform; avoid misunderstanding by asking just what the benchmark is measuring. *IEEE Spectrum*, 24(7):38–43, 1987.

[31] J. J. Dongarra, C. B. Moler, J. R. Bunch, and G. W. Stewart. *LINPACK users' guide.* SIAM, 1979.

[32] K. Ebrahimi, G. F. Jones, and A. S. Fleischer. A review of data center cooling technology, operating conditions and the corresponding low-grade waste heat recovery opportunities. *Renewable and sustainable energy reviews*, 31:622–638, 2014.

[33] European Green Digital Coalition. Net Carbon Impact Assessment Methodology for ICT Solutions. `https://www.greendigitalcoalition.eu/assets/uploads/2024/04/EGDC-Net-Carbon-Impact-Assessment-Methodology-for-ICT-Solutions.pdf`, 2024. Accessed: (June 2025).

[34] D. G. Feitelson, L. Rudolph, U. Schwiegelshohn, K. C. Sevcik, and P. Wong. Theory and practice in parallel job scheduling. In *Job Scheduling Strategies for Parallel Processing: IPPS'97 Processing Workshop*

Geneva, Switzerland, April 5, 1997 Proceedings 3, pages 1–34. Springer, 1997.

[35] F. Fraternali, A. Bartolini, C. Cavazzoni, and L. Benini. Quantifying the Impact of Variability and Heterogeneity on the Energy Efficiency for a Next-Gen. Ultra-Green Supercomputer. *IEEE Trans. on Parallel and Dist. Systems*, 29(7), 2018.

[36] Y. Gao and P. Zhang. A Survey of Homogeneous and Heterogeneous System Architectures in High Performance Computing. In *IEEE International Conf. on Smart Cloud*, pages 170–175, 2016.

[37] H. Geller, P. Harrington, A. H. Rosenfeld, S. Tanishima, and F. Unander. Polices for increasing energy efficiency: Thirty years of experience in oecd countries. *Energy policy*, 34(5):556–573, 2006.

[38] M. Ghamkhari, A. Wierman, and H. Mohsenian-Rad. Energy portfolio optimization of data centers. *IEEE Transactions on Smart Grid*, 8(4):1898–1910, July 2017.

[39] M. Ghasemi-Gol, Y. Wang, and M. Pedram. An optimization framework for data centers to minimize electric bill under day-ahead dynamic energy prices while providing regulation services. In *Int. Green Comp. Conf.*, pages 1–9, Nov 2014.

[40] J. Giordani, Z. Xu, E. Colby, A. Ning, B. R. Godala, I. Chaturvedi, S. Zhu, Y. Chon, G. Chan, Z. Tan, et al. Revisiting computation for research: Practices and trends. In *SC24: International Conference for High Performance Computing, Networking, Storage and Analysis*, pages 1–14. IEEE, 2024.

[41] W. E. Gnibga, A. Blavette, and A.-C. Orgerie. Renewable energy in data centers: the dilemma of electrical grid dependency and autonomy costs. *IEEE Transactions on Sustainable Computing*, 9(3):315–328, 2023.

[42] W. E. Gnibga, A. A. Chien, A. Blavette, and A. C. Orgerie. Flexcooldc: datacenter cooling flexibility for harmonizing water, energy, carbon, and cost trade-offs. In *Proceedings of the 15th ACM International Conference on Future and Sustainable Energy Systems*, pages 108–122, 2024.

[43] Í. Goiri, K. Le, M. E. Haque, R. Beauchea, T. D. Nguyen, J. Guitart, J. Torres, and R. Bianchini. Greenslot: scheduling energy consumption in green datacenters. In *Proceedings of 2011 International Conference for High Performance Computing, Networking, Storage and Analysis*, pages 1–11, 2011.

[44] B. Gomes and K. Brandt. Our 2024 Environmental Report, July 2024. `https://blog.google/outreach-initiatives/sustainability/2024-environmental-report/`. Accessed: (April 2025).

[45] M. Gregory. Inside Facebook's green and clean arctic data centre. BBC News, Business, June 2013. https://www.bbc.com/news/business-22879160. Accessed: (April 2025).

[46] U. Gupta, M. Elgamal, G. Hills, G.-Y. Wei, H.-H. S. Lee, D. Brooks, and C.-J. Wu. Act: Designing sustainable computer systems with an architectural carbon modeling tool. In *Proceedings of the 49th Annual International Symposium on Computer Architecture*, pages 784–799, 2022.

[47] M. Haddad, G. Da Costa, J.-M. Nicod, M.-C. Péra, J.-M. Pierson, V. Rehn-Sonigo, P. Stolf, and C. Varnier. Combined IT and power supply infrastructure sizing for standalone green data centers. *Sustainable Computing: Informatics and Systems*, 30:100505, 2021.

[48] M. Haddad, J.-M. Nicod, M.-C. Péra, and C. Varnier. Stand-alone renewable power system scheduling for a green data center using integer linear programming. *Journal of Scheduling*, 24:523–541, 2021.

[49] L. M. Hilty, P. Arnfalk, L. Erdmann, J. Goodman, M. Lehmann, and P. A. Wäger. The relevance of information and communication technologies for environmental sustainability–a prospective simulation study. *Environmental Modelling & Software*, 21(11):1618–1629, 2006.

[50] C.-H. Hsu and W.-C. Feng. The Right Metric for Efficient Supercomputing: A Ten-Year Retrospective. In *2016 IEEE International Parallel and Distributed Processing Symposium Workshops (IPDPSW)*, pages 1090–1093, 2016.

[51] Z.-H. Hu, Y.-X. Zheng, and Y.-G. Wang. Packing computing servers into the vessel of an underwater data center considering cooling efficiency. *Applied Energy*, 314:118986, 2022.

[52] U. Hölzle. 100% renewable is just the beginning, 2016. https://sustainability.google/projects/announcement-100/. Accessed: (April 2025).

[53] IBM. Perf, 2025. https://perfwiki.github.io/main/, 2025-04-22.

[54] IEA. Energy and AI. Technical report, International Energy Agency, Paris, 2025. https://www.iea.org/reports/energy-and-ai, Licence: CC BY 4.0.

[55] Intergovernmental Panel on Climate Change (IPCC). *Global Warming of 1.5°C*, page 1–24. Cambridge University Press, 2022.

[56] M. Jay, V. Ostapenco, L. Lefèvre, D. Trystram, A.-C. Orgerie, and B. Fichel. An experimental comparison of software-based power meters: focus on cpu and gpu. In *2023 IEEE/ACM 23rd International Symposium on Cluster, Cloud and Internet Computing (CCGrid)*, pages 106–118. IEEE, 2023.

[57] W. S. Jevons. *The coal question; an inquiry concerning the progress of the nation and the probable exhaustion of our coal-mines.* Macmillan, 1866.

[58] L. H. Kaack, P. L. Donti, E. Strubell, G. Kamiya, F. Creutzig, and D. Rolnick. Aligning artificial intelligence with climate change mitigation. *Nature Climate Change*, 12(6):518–527, 2022.

[59] A. Kassab, J.-M. Nicod, L. Philippe, and V. Rehn-Sonigo. Green power aware approaches for scheduling independent tasks on a multi-core machine. *Sustainable Computing: Informatics and Systems*, 31:100590, 2021.

[60] A. Khan, H. Sim, S. S. Vazhkudai, A. R. Butt, and Y. Kim. An Analysis of System Balance and Architectural Trends Based on Top500 Supercomputers. In *The International Conference on HPC in Asia-Pacific Region*, (HPC Asia). ACM, 2021.

[61] P. M. Kogge and T. J. Dysart. Using the TOP500 to Trace and Project Technology and Architecture Trends. In *Proceedings of 2011 International Conference for HPC, Networking, Storage and Analysis*, pages 1–11, 2011.

[62] J. Koomey, Z. Schmidt, and S. Naffziger. Supercomputing performance and efficiency: an exploration of recent history and near-term projections. *Burlingame, CA: Koomey Analytics, for AMD*, 2020.

[63] D. Landré, J.-M. Nicod, and C. Varnier. Optimal standalone data center renewable power supply using an offline optimization approach. *Sustainable Computing: Informatics and Systems*, 34:100627, 2022.

[64] LAZARD. Lazard's levelized cost of storage analysis—version 7.0, 2021. `https://www.lazard.com/media/42dnsswd/lazards-levelized-cost-of-storage-version-70-vf.pdf`. Accessed: (April 2025).

[65] A.-L. Ligozat and A. D. Vries. Generative AI: Energy consumption soars, Nov. 2024. `https://www.polytechnique-insights.com/en/columns/energy/generative-ai-energy-consumption-soars`. Accessed: (April 2025).

[66] Q. Liu and Z. Yu. The Elasticity and Plasticity in Semi-Containerized Co-locating Cloud Workload: a View from Alibaba Trace. In *Proceedings of the ACM Symposium on Cloud Computing*, SoCC '18, page 347–360, New York, NY, USA, 2018. Association for Computing Machinery.

[67] F. Lossin, I. Kozlovskiy, M. Sodenkamp, and T. Staake. Incentives to go green: an empirical investigation of monetary and symbolic rewards to motivate energy savings. In *ECIS Proceedings*, 2016.

[68] A. S. Luccioni, E. Strubell, and K. Crawford. From Efficiency Gains to Rebound Effects: The Problem of Jevons' Paradox in AI's Polarized Environmental Debate. *arXiv preprint arXiv:2501.16548*, 2025.

[69] J. Mair, Z. Huang, D. Eyers, and Y. Chen. Quantifying the Energy Efficiency Challenges of Achieving Exascale Computing. In *15th IEEE/ACM International Symposium on Cluster, Cloud and Grid Computing*, pages 943–950, 2015.

[70] J. Malmodin, N. Lövehagen, P. Bergmark, and D. Lundén. ICT sector electricity consumption and greenhouse gas emissions–2020 outcome. *Telecommunications Policy*, 48(3):102701, 2024.

[71] T. Menouer, C. Cérin, and P. Darmon. Koptim: Kubernetes optimization framework. In *IEEE International Parallel and Distributed Processing Symposium, IPDPS 2024 Workshop, San Francisco, CA, USA, May 27-31, 2024*, pages 900–908. IEEE, 2024.

[72] T. Menouer, C. Cérin, and P. Darmon. Reactive autoscaling of kubernetes nodes. In M. Coppola, H. Kavalionak, L. Ferrucci, and I. Kontopoulos, editors, *Proceedings of the 4th Workshop on Flexible Resource and Application Management on the Edge, FRAME 2024, Pisa, Italy, June 3-7, 2024*. ACM, 2024.

[73] T. Menouer, C. Cérin, P. Darmon, and J. Rivalan. Dynamic configuration of kubernetes containers resources with sla classes. In *IEEE International Parallel and Distributed Processing Symposium, IPDPS 2025 PDCO Workshop, Milan, Italy, June 3-8, 2025*. IEEE, 2025.

[74] T. A. Norton, S. L. Parker, M. C. Davis, S. V. Russell, and N. M. Ashkanasy. A virtuous cycle: How green companies grow green employees (and vice versa). In *Research handbook on employee pro-environmental behaviour*, pages 210–228. Edward Elgar Publishing, 2018.

[75] NVIDIA Corporation. Nvidia management library (nvml), 2025. `https://developer.nvidia.com/management-library-nvml`. Accessed: (April 2025).

[76] ORNL. Frontier's architecture. `https://www.olcf.ornl.gov/wp-content/uploads/Frontier-Architecture-Overview\_Abraham.pdf`, 2024. Accessed: (October 2024).

[77] E. L. Perasso, C. Vateau, F. Domon, et al. Evaluation de l'impact environnemental du numérique en france et analyse prospective. Technical report, Technical report, ADEME, 2022.

[78] J.-M. Pierson, G. Baudic, S. Caux, B. Celik, G. Da Costa, L. Grange, M. Haddad, J. Lecuivre, J.-M. Nicod, L. Philippe, et al. Datazero:

Datacenter with zero emission and robust management using renewable energy. *Ieee Access*, 7:103209–103230, 2019.

[79] T. Pirson, T. Delhaye, A. Pip, G. Le Brun, J.-P. Raskin, and D. Bol. The environmental footprint of ic production: Meta-analysis and historical trends. In *ESSDERC 2022-IEEE 52nd European Solid-State Device Research Conference (ESSDERC)*, pages 352–355. IEEE, 2022.

[80] G. Rostirolla, L. Grange, T. Minh-Thuyen, P. Stolf, J.-M. Pierson, G. Da Costa, G. Baudic, M. Haddad, A. Kassab, J.-M. Nicod, et al. A survey of challenges and solutions for the integration of renewable energy in datacenters. *Renewable and Sustainable Energy Reviews*, 155:111787, 2022.

[81] R. Schwartz, J. Dodge, N. A. Smith, and O. Etzioni. Green AI. *Communications of the ACM*, 63(12):54–63, 2020.

[82] T. Scogland, B. Subramaniam, and W.-C. Feng. Emerging Trends on the Evolving Green500: Year Three. In *IEEE International Symposium on Parallel and Distributed Processing Workshops and Phd Forum*, pages 822–828, 2011.

[83] A. Shehabi, A. Hubbard, A. Newkirk, N. Lei, M. A. B. Siddik, B. Holecek, J. Koomey, E. Masanet, D. Sartor, et al. 2024 united states data center energy usage report. Research report LBNL-2001637, Lawrence Berkeley National Laboratory, Berkeley, California, 2024.

[84] A. L. V. Solórzano, K. Sato, K. Yamamoto, F. Shoji, J. M. Brandt, B. Schwaller, S. P. Walton, J. Green, and D. Tiwari. Toward Sustainable HPC: In-Production Deployment of Incentive-Based Power Efficiency Mechanism on the Fugaku Supercomputer. In *SC24: International Conference for High Performance Computing, Networking, Storage and Analysis*, pages 1–16. IEEE, 2024.

[85] T. Spencer and S. Singh. What the data centre and AI boom could mean for the energy sector. *International Energy Agency (IEA)*, 7:2025, 2024.

[86] E. Strohmaier, H. W. Meuer, J. Dongarra, and H. D. Simon. The TOP500 List and Progress in High-Performance Computing. *Computer*, 48(11):42–49, 2015.

[87] E. Suarez, J. Amaya, M. Frank, O. Freyermuth, M. Girone, B. Kostrzewa, and S. Pfalzner. Energy Efficiency trends in HPC: What high-energy and astrophysicists need to know. *arXiv preprint arXiv:2503.17283*, 2025.

[88] B. Subramaniam, W. Saunders, T. Scogland, and W.-C. Feng. Trends in energy-efficient computing: A perspective from the Green500. In

2013 International Green Computing Conference Proceedings, pages 1–8, 2013.

[89] T. Tashev. 12 Biggest Data Centers in the World, 2024. `https://www.webopedia.com/technology/10-biggest-data-centers-in-the-world`. Accessed: (April 2025).

[90] P. Taylor. Data center average annual Power Usage Effectiveness (PUE) worldwide from 2007 to 2024, Apr. 2025. `https://www.statista.com/statistics/1229367/data-center-average-annual-pue-worldwide`. Accessed: (April 2025).

[91] M.-T. Thi, J.-M. Pierson, G. Da Costa, P. Stolf, J.-M. Nicod, G. Rostirolla, and M. Haddad. Negotiation game for joint IT and energy management in green datacenters. *Future Generation Computer Systems*, 110:1116–1138, 2020.

[92] J. Treibig, G. Hager, and G. Wellein. Likwid: A lightweight performance-oriented tool suite for x86 multicore environments. In *2010 39th International Conference on Parallel Processing workshops*, pages 207–216. IEEE, 2010.

[93] R. Watt. The fantasy of carbon offsetting. *Environmental Politics*, 30(7):1069–1088, 2021.

[94] V. M. Weaver, M. Johnson, K. Kasichayanula, J. Ralph, P. Luszczek, D. Terpstra, and S. Moore. Measuring energy and power with PAPI. In *2012 41st international conference on parallel processing workshops*, pages 262–268. IEEE, 2012.

[95] Wikipedia. 2021 Texas power crisis. `https://en.wikipedia.org/wiki/2021\_Texas\_power\_crisis`, 2023. Accessed: (Nov. 2023).

[96] World Resources Institute. Do we need a standard to calculate "avoided emissions"? `https://www.wri.org/insights/do-we-need-standard-calculate-avoided-emissions`, 2013. Accessed: (June 2025).

[97] A. B. Yoo, M. A. Jette, and M. Grondona. Slurm: Simple linux utility for resource management. In *JSSPP 2003*, pages 44–60. Springer, 2003.

[98] X. Yuan, X. Zhou, Y. Pan, R. Kosonen, H. Cai, Y. Gao, and Y. Wang. Phase change cooling in data centers: A review. *Energy and Buildings*, 236:110764, 2021.

[99] Q. Zhang, Z. Meng, X. Hong, Y. Zhan, J. Liu, J. Dong, T. Bai, J. Niu, and M. J. Deen. A survey on data center cooling systems: Technology, power consumption modeling and control strategy optimization. *Journal of Systems Architecture*, 119:102253, 2021.

[100] Z. Zhang, C. Li, Y. Tao, R. Yang, H. Tang, and J. Xu. Fuxi: a fault-tolerant resource management and job scheduling system at internet scale. *Proc. VLDB Endow.*, 7(13):1393–1404, Aug. 2014.

[101] D. Zhao, S. Samsi, J. McDonald, B. Li, D. Bestor, M. Jones, D. Tiwari, and V. Gadepally. Sustainable supercomputing for AI: GPU power capping at HPC scale. In *Proceedings of the 2023 ACM Symposium on Cloud Computing*, pages 588–596, 2023.

Index

For Product Safety Concerns and Information please contact our
EU representative GPSR@taylorandfrancis.com Taylor & Francis
Verlag GmbH, Kaufingerstraße 24, 80331 München, Germany